DOES SCIENCE NEED A GLOBAL LANGUAGE?

English and the Future of Research

Scott L. Montgomery

With a Foreword by David Crystal

THE UNIVERSITY OF CHICAGO PRESS * *Chicago and London*

SCOTT L. MONTGOMERY is a consulting geologist and independent scholar, and the author of numerous books including *The Chicago Guide to Communicating Science* and *Science in Translation*, both published by the University of Chicago Press.

The University of Chicago Press, Chicago 60637
The University of Chicago Press, Ltd., London

Printed in the United States of America

22 21 20 19 18 17 16 15 14 13 1 2 3 4 5

ISBN-13: 978-0-226-53503-6 (cloth)
ISBN-13: 978-0-226-01004-5 (e-book)

Library of Congress Cataloging-in-Publication Data

Montgomery, Scott L.
Does science need a global language? : English and the future of research / Scott L. Montgomery ; with a foreword by David Crystal.
pages ; cm
Includes bibliographical references and index.
ISBN 978-0-226-53503-6 (hardcover : alkaline paper)—
ISBN 978-0-226-01004-5 (e-book) 1. Science—Language. 2. English language—Social aspects. I. Crystal, David, 1941- II. Title.
Q226.M658 2013
501′.4—dc23

2012027704

♾ This paper meets the requirements of ANSI/NISO Z39.48-1992 (Permanence of Paper).

TO

Russ Sutton and

William B. Travers,

who helped start me on the path

CONTENTS

Foreword by David Crystal ix

Preface xi

(1)

A New Era

1

(2)

Global English

Realities, Geopolitics, Issues

24

(3)

English and Science

The Current Landscape

68

(4)

Impacts

A Discussion of Limitations and Issues for a Global Language

102

(5)
Past and Future
What Do Former Lingua Francas of Science Tell Us?
132

(6)
Does Science Need a Global Language?
166

Notes 189 * *Index* 215

FOREWORD

The first books analyzing the phenomenon of English as a global language appeared during the late 1990s, and they inevitably adopted a general perspective. There was a concern to identify all the factors relevant to explaining why English had become global, and what had happened to the language as a result. In the absence of large-scale surveys or empirical case studies, writers were forced to be personal, anecdotal, and at times superficial. I know, because I was one of them. Looking back at my *English as a Global Language* (1997), I am struck now by how many of its observations lacked detail. I identified a wide range of situations demonstrating global presence, such as the role of English in advertising, broadcasting, popular music, and science; but an introductory few hundred words on each topic could do no more than point to a series of stories that needed to be told in the next generation of studies exploring how English is actually used in these domains.

Scott Montgomery's book is a fine illustration of the way these second-generation studies are emerging. It is the scientific story, told by a scientist—but a scientist who has taken the trouble to understand current linguistic thinking on the global use of English. "English is the language of science" is one of the most commonly encountered claims, and cries out for the kind of detailed examination that he now provides. How recent is the adoption of English by scientists? Is English used to the same extent in all scientific domains? How are scientists who use English responding to the emergence of international variation as "new Englishes" present users with competing standards? What happens linguistically when the editorial management of scientific journals shifts from being exclusively Anglo-American to being multinational? What counts as acceptable scientific English in a world of new Englishes? These and numerous other such questions are all explored with perceptive insight in this friendly and personal book.

As Montgomery writes in his preface, "What has happened to modern science is remarkable, revolutionary." From a linguistic point of view, I already knew it was remarkable, for the concept of a single lingua franca in science is itself an extraordinary development. But I had not fully taken on board just how revolutionary the consequences of this development have been. Montgomery continues, "It is this ability of scientists throughout the world to speak and write to each other, to read each others' work directly, and to collaborate without mediators of any kind that defines the new era." A new era. I had not thought of it like this before. I am used to reflecting on the fact that for every native speaker of English there are now four or five nonnative English speakers in the world. The center of gravity in the language has shifted, and English is changing its character as a result. But I had not reflected, until now, on the implications of this for the language of science. The center of gravity has shifted here, too, as Scott Montgomery makes very clear, and consequently scientific English has begun to change.

An inevitable naïveté in English language studies comes from linguists viewing the domains in which the language is used without fully appreciating the complexity of the issues they contain. To truly appreciate the role of English in one of these domains, we have to live it. The most perceptive studies on the language of advertising, for example, are those in which the linguist-writer has at some time been an advertiser, or at least collaborated with one. And it is the same with science. Either English language experts learn as much as they can about science, or scientists learn as much as they can about English. I have tried to do the former, aided by the fact that in many of its dealings, linguistics already is a science. Scott Montgomery has tried to do the latter. And it is this complementarity of interests that makes me especially pleased to write the foreword to this book.

David Crystal

PREFACE

A few words to the reader will be helpful. Like thousands of scientists, I have been in contact with researchers from many parts of the world over the years, writing collaborative research papers, giving talks at large meetings, serving on panels, consulting for major companies, acting as a journal reviewer and editor. I've engaged in less common but related activities, too, as a translator of scientific material and as a historian of science and student of scientific language. I have watched this language evolve over the last three decades since I began to write for publication. Yet the most striking change has not been in the character or style of scientific discourse.

In 1977, as a grad student, I was assigned to work with a visiting professor from Iran who needed to learn about plate tectonics. It was a difficult, frustrating, and ultimately unsuccessful assignment, because we knew little of each other's languages. He spoke both Farsi and Arabic, but almost no English. For weeks, we could not communicate. Eventually, we created our own pidgin with sometimes curious results (the phrase "is the car?" was what I came up with as the Arabic equivalent for "convergence"). Thirty years later, during the summer of 2011, I gave a lecture on science and language to a group of visiting engineering students from Iraq, Oman, and Yemen, and was peppered with questions in English about Arabic as an international scientific tongue. One Iraqi student, a mechanical engineer, offered an impassioned soliloquy about how Baghdad had been the center of the intellectual world, not least in medicine ("They practiced actual dissection"). He grew quiet for a moment, then said, "The Mongols destroyed it all. It ended Islam's golden age in science."

What has happened to modern science is remarkable, revolutionary. Many big changes are interpreted for the sciences—the Internet, with its new forms of networking; international projects of unprecedented scale, such as the Large Hadron Collider; major leaps in the knowledge of certain fields, such as the decoding of the human genome. Yet, at a certain level, all of these come back to language. They depend directly on a shared ability to communicate, on a global tongue. It is this ability of scientists throughout the world to speak and write to each other, to read each other's work directly, and to collaborate without mediators of any kind that defines the new era. It is an era in which more science can be done more quickly in more places than ever before.

My use of the term *science*, it must be said, is a bit encompassing, including as it does the natural sciences, medicine, and a good deal of engineering—what most researchers would agree to label as scientific disciplines in terms of basic outlook, method, and training. Indeed, I have spoken about the subject of this book with many natural scientists, physicians, and engineers, to gain insight from their opinions and, perhaps even more, their own life stories. Language is nothing if not an intensely human reality of exchange. It seemed only fitting that I include a number of these stories—mainly from the present day, but a few from other times—to help suggest the many levels of experience that are involved in a global language for science, the advantages and the limits such a language brings, and the greater historical and even geopolitical circumstances involved. For reasons of privacy, I have changed the names in most cases.

* * *

The debts I have acquired in writing this book are great and my ability to repay them trifling. Always to be remembered, with appreciation, are the many scientists, physicians, engineers, and university students, from many parts of the world, who have shared their time with me over the past decade. A simple list of these names, if I could assemble it, would run for many pages. I choose, therefore, to thank them all at once, as evidence of my incapacity to recompense their kindness. A few individuals do merit special note, however. Robert Winglee at the University of Washington, Earth and Space Sciences, deserves my thanks for allowing me to accompany his group to Northern Australia in 2010, giving me the chance to discuss language issues with a number of people from that part of the world. Thanks go as well to Professor Steve Harrall for his recommendations on source material and his invitation to visit the Yakima Indian reservation with his class. Input on certain questions from David Bellos and Michael Gordin at Princeton University and Maeve Olohan at the University of Manchester also proved helpful and is much appreciated. Finally, my thanks go to David Crystal for his encouragement and kind support.

None of these people can be held responsible for any errors or mistakes of judgment that might occur in this work; they are all my own.

As always, too, one must live to write. Life with Marilyn, Kyle, and Cameron, and now Clio, has made it possible for me to complete yet another effort of this kind.

Adeste fideles

(1)

A New Era

To peep at such a world—to see the stir
Of the great Babel . . .
WILLIAM COWPER, *"The Task"*

When I first met Ben, I thought he must be in customer service, so friendly and practiced was his smile. In fact, he is a biochemist from Uganda. Very dark-skinned and always neatly dressed with a touch of elegance, he speaks fluent and natural English that bubbles with an East African accent. His eyes have a sharp intelligence that can penetrate solid objects. We were forced colleagues, our boys playing on the same sports team, and so I decided to ask how he became a chemist. Every researcher has a story; his was something more.

"I am lucky to be a scientist," Ben began, "but my luck was no accident." Born in 1958, four years before Uganda's independence from Britain, Ben spent his early years near Mubende, a town northwest of Kampala, where Bantu is spoken. He attended a local district school like most other children and was taught English. His father had worked in the colonial bureaucracy and often spoke the language at home with his son. "He had high hopes for me," Ben said, without further explanation. "He saved enough to send me to a private academy, where a British man taught." This man, an expatriate engineer of Indian descent, quickly recognized in Ben an aptitude for math. With the father's permission, he gave the boy private lessons and much encouragement. "He was a mentor," Ben noted, "and a lifeline."

In 1972, the new dictator, Idi Amin, ordered all Asians to leave the country within ninety days. The teacher was forced to flee and never returned. In the face of mounting chaos and murders caused by the regime, Ben's father sent his son to an uncle in Tanzania and then, with help from other family members, to San Francisco, where a relative owned a small restaurant. Ben was granted refugee status and attended school while working part time in the restaurant; since his English was both excellent and polite, he helped conduct business with suppliers. With his earnings, he eventually enrolled in a community college. Ben's parents told him he must remain in the United States, so he eventually transferred to the University of Oregon, where a scholarship helped him earn a BS in mathematics and an MS in biochemistry. Chemistry drew him, he said, because of its powers of transformation. "I know this is the ancient view, of the alchemists. But it is true; in chemistry I found a kind of hope." He studied the biochemistry of plants for his PhD, then took a job with a firm in Chicago.

Since 1990, Ben has specialized in food-related research. When I asked why, he replied, "Because this is what the world needs most." He has had professional assignments in Brazil, India, Japan, Norway, and elsewhere, and has presented papers at many scientific meetings. He enjoys these meetings a great deal and attends several every year, as he almost always comes away with new research ideas and collaborations. Yet he said he had been thinking about returning to Uganda to teach. When I expressed surprise at this desire to end a successful career, he looked at me without smiling. "I feel science must be shared," he said. "It is not mine to keep. I can speak to my countrymen in a language that will not take sides with any group."

Science, Globally Speaking

In a globalizing world, language is power. The more human beings and institutions with which we can communicate, the more access to the offerings and agents of the larger world we gain. This may seem merely a matter of numbers, but far more is involved. Language has a role in the oldest dream for a better world: the dream of a universal language that allows people everywhere to commune and work together. It is the vision of a unified humanity, harmony on a planetary scale. In the West, we know this dream through the image of its loss: the biblical story of the Tower of Babel, a great structure erected to reach the heavens, designed no doubt by engineers and scientists of the time, but left incomplete when a jealous God shattered the once-universal language into thousands of tongues that could not understand one another.

What if, after a significant pause, a new chapter and verse might be added to the tale? What if, in our own time, a worthy alternative to Babel has emerged, lacking in arrogance, extending not merely to the empyreal realm but deep into the atom and as far as the distant galaxies? Such questions have already been answered. For the first time in history, science—humanity's great tower of knowledge—has a global tongue. In truth, it is a global language for numerous domains, with science being one case among many. It is a special case, to be sure, but one whose meaning can't be probed without an understanding of this larger reality.

Today, close to 2 billion people in over 120 nations speak English at some level of proficiency.[1] This extraordinary number includes a broad spectrum of ability, without any doubt. Yet it testifies to the global draw this language now commands. For the natural sciences, medicine, and large areas of engineering, English utterly dominates in international communication. This does not mean that it rules in every circumstance, in every country. Its dominance has definite limits, being confined mainly to situations with an international or, especially, a global dimension. Yet this is crucial, as we will see, since science has itself entered a new, globalizing era. English, in short, is the global tongue for this era of globalization.

By the late 2000s, nearly all forms of written output, whether in print or online, whether in person or in video, whether in professional or informal settings, had already come to depend on this one tongue when the intended audience is the larger world community of researchers in any field. Scientists everywhere now recognize this. They would find it necessary to also stress that the global role of this language isn't at all confined to publication. English has become the speech of international scientific conferences, symposia, conventions, colloquia, visiting lectures, workshops, interviews, and more—the oral dimension to global science. When Ben goes to Brazil or Japan to give a three-week minicourse on protein synthesis in dwarf wheat, or when he is hired as a consultant by a German agricultural firm to examine its operations in Southeast Asia, he speaks English. As he explained, this isn't an accommodation by his clients but a requirement, a company policy. "I would not be hired for these jobs unless I spoke it," he said.

Corporate scientific exchanges, whether between European and African firms or among Asian companies from different countries, also rely on English. Indeed, private sector science led by multinational firms that invest in research and development (R&D), training, and new facilities depend on this language. International patents are now overwhelmingly filed in English. Online postings of research jobs, postdoctoral fellowships, new databases and other resources, and international grants all now employ English as well.

Then there is scientific information itself. Websites of major research institutes, organizations, and statistical and data archives around the globe—core repositories of contemporary technical knowledge—have turned to English. A tiny sampling of these might include CERN (Conseil Européen pour la Recherche Nucléaire; now European Organization for Nuclear Research), PubMed (largest archive worldwide for the biomedical and life sciences), ChemWeb (for chemistry), GeoRef (for earth sciences), ENCODE (data from the human genome sequence), OBIS (Ocean Biogeographic Information System), arXiv (preprint archive for physics, mathematics, and other fields), the Max Planck Institute, the European Science Foundation, the United Nations Statistical Databases, and the global Census of Marine Life. For Internet science in general, we need only use a term such as *natrium* (the Dutch word for "sodium") or *RNA* as a search word to witness the paucity of sites retrieved that appear in any language other than English. Of course, Internet search engines capture only what has been most often used—but that's exactly the point.

Will science conducted in other languages die out before long? Not at all. Throughout the world, many thousands of technical journals are published in Chinese, Japanese, Portuguese, Russian, French, Spanish, Korean, Arabic, and so on. Despite the growth in the use of English in scientific endeavors, there is little likelihood that domestic science—fortified by the demands of competitive nationalism and *realpolitik*—will go away any time soon. Governments fund science less for love of truth than for economic competitiveness, defense, prestige, public health. If such aims remain firm, and if scientific work looks to the embrace of government support in its homeland, a healthy national literature will continue. In a few countries where English is close to being a second language (Scandinavia, for example), it is true that the native tongues are much less used even in domestic scientific communication. On the other hand, in other regions, such as Latin America, the domestic tongue is also a world language. Thus, if English seems an overwhelming force in some places, it is much less so in others. By no measure does it command a true hegemony. Again, its realm has limits. Where it remains unrivaled is in science's expanding global dimension: the greater internationalization of new knowledge and its creation.

At first blush, a new linguistic competitor does seem to have emerged. Mandarin Chinese, with roughly 900 million users and backed by China's own spectacular economic rise, is felt by many to be capable of replacing English in science and elsewhere over the next several decades or so. Between 1999 and 2009, the annual number of scientific publications that included one or more Chinese authors increased from less than 30,000 to nearly 120,000 in international journals—a fourfold leap in a single decade (by comparison, US output grew only 30%, from 265,000 to 340,000).[2] Furthermore, it has become common to hear Chinese spoken in the hallways of science and engineering graduate departments all across the United States. There are numbers to back this up: by 2009, foreign students, especially those from China and India, earned no less than 33% of all doctoral degrees granted by US institutions in the sciences and 57% in engineering.[3] Such are figures to give one pause.

But to think that they reveal a new tide of favor for the Chinese language over English would be naïve. Impressive as the publication statistics surely are, what they show is the success of Chinese researchers in *English*—the language of international journals in science—not Mandarin. The speed with which Chinese representation has risen in measured publications reflects directly how rapidly English has been accepted by Chinese scientists as dominant in the global context. As for Chinese students in America, it would be an error to think of them as "agents" working on behalf of their native tongue. Even the most informal survey will show that their goals include gaining a higher level of scientific training and improved English-language skills, not least a stronger ability to write and publish in this language.* By far the largest source of funding for these students are their own families, not the government (they are not linguistic infiltrators!). Ask them the language in which they wish to publish their research, and you will find a single answer. "If we want a research job here or in Europe, or in an international company, or even a high level job back in China, too, we *must* publish in English," a physics PhD candidate told me recently. "China's best scientists do this. They want international audiences, and this means English." Recognizing this truth, most of China's top research journals — over two hundred by 2010 — are themselves changing to English-only publication.[4] Most major research institutions in China, not least the Chinese Academy of Sciences, now have versions of their websites, journals, papers, and databases in English. Dozens of major Chinese universities offer science and engineering courses in English, to both foreign and Chinese students. One would be hard pressed to find even a single institution in North America or Europe following this course for Chinese. Meanwhile, English-language courses in the sciences now appear in the curricula of universities worldwide, from Finland to Korea.

We can approach this phenomenon from a different angle. It is estimated that over 1.5 billion people worldwide, including schoolchildren, are learning English to varying degrees, while about 30 million to 40 million can be counted as studying Mandarin, with a far smaller number (in the low thousands) learning Cantonese and other Chinese languages.[5] Thus, the number of Chinese learners would need to grow by about forty times to compete with English at a significant level. According to a speech given by Chinese premier Wen Jiabao in 2009, more than 300 million of his countrymen were studying English that year, compared with about 50,000 Americans learning Chinese. This figure for Chinese learners of English is likely exaggerated; yet common estimates by those less engaged

*During the course of writing this book, I have spoken with over 150 Chinese graduate students and postdocs at US and Canadian universities, plus another 50 or so researchers in the corporate sector. Over 90% of these individuals mentioned either improving their English skills or publishing papers in English as being among their major professional aims. Many of them also expressed the wish that Westerners would learn Chinese too.

in political speech making begin at 100 million to 180 million and go up from there.[6] Then there is the fact that English has become a required subject in Chinese schools starting in grade 3; only 4% of US middle and high schools were even offering Mandarin as a choice by 2008–9, when the news media spoke of a "great surge" in interest for this language.[7] An important work on the status of English in China published in 2009 made the overall point even more forcefully.

> It is clear that English learning is unlike the teaching and learning of other FLs [foreign languages] in ways beyond issues of scale or size. . . . China is regarded as an EFL [English as a Foreign Language] country, but the depth of penetration and the variegated roles assigned to English, the "local reward systems" available through English, point to levels of domestication more typical of English as a Second Language settings. This implies that English is imagined to have, and Chinese society has taken steps to bring about, *domestic social functions* for English: English for Chinese purposes in Chinese settings. In a key, if limited sense, this aims to make English a Chinese language.[8]

None of this is to say that the situation is final. Major changes can certainly occur during the present century. Yet any such changes would have to reverse a momentum of profound, global extent. Empirically, the dominance of English in science stands beyond question. From lab to classroom, democracy to autocracy, researchers can and do communicate well in a language accepted as a kind of universal currency.

It would be wrong, however, to assume that scientists everywhere possess this coin, or possess it to the same degree. They do not. And as with any form of capital, uneven possession is widespread and means inequality, with large implications. There are realities that a story like Ben's doesn't bring to the eye or ear. An international tongue can be a hard master. Those who have it, as Ben did from an early age ("my luck was not an accident"), may gain opportunity, mobility, and more. But consider the young Korean biochemist whose English is poor, who must struggle or pay to get her slides translated for an upcoming meeting, to work on her script, pronunciation, anxiety. Those who do not possess command of the dominant tongue find themselves limited, confined, even disenfranchised, ignored. Much forced accommodation exists among scientists who do not know English well. Local tongues and possibly cultures are affected. A language that spreads to many nations is one toward which many millions of people will migrate, perhaps leaving behind part of their native linguistic heritage. Casualties exist, in other words. History (as we will see in chapter 5) suggests that they may not be avoidable.

Yet there are huge advantages to learning the dominant tongue, another point taken from history. This is of course a reason why so many scientists and engineers do so without a sense of injustice. They place great value on being able to reach so many others, to gain higher levels of recognition for their work, to join and share this work with the greatest number. If we look at the past, we see this has happened time and again. A language such as Latin or Arabic began its spread as the tongue of conquerors and traders, but then evolved into an unparalleled storehouse of texts and knowledge that remained vital long after the respective empires crumbled away. Few scholars of nature in sixteenth-century Paris or tenth-century Cordoba could have made a contribution without Renaissance Latin or classical Arabic. These languages provided reservoirs into which thinkers from many places added their contributions. As languages of power, they weakened the motive for doing sophisticated work in local tongues. But this did not mean that thinkers had to abandon their speech as a matter of course. The situation has rarely been either/or. For researchers today as well, knowledge of English is often an added skill.

The Question

So science has a global language. Does it *need* such a tongue? Do the benefits truly outweigh any limitations or problems? And what of the future? Will science gain or lose vitality, invention, creativity, by having such a language?

Here is the main matter. A simple yes or no won't suffice—not without a good deal of explanation and, perhaps, qualifying details. As with all things linguistic, the matter is messy. English in science is a subset of English in the world. The first can't be understood without some understanding of the causes and impacts, therefore the issues, that have emerged as a result of the second. I will reserve an answer to our question until later (though with many hints at it along the way).

Doubtless many scientists would simply answer in the affirmative and beg leave to return to the lab. Yet there are others who, while certainly recognizing it, nonetheless bemoan the status and power of English today for some of the reasons just mentioned. This must be acknowledged. A global tongue raises ethical issues and disputes to a higher level. For instance, the endangerment of native languages throughout the world, now well documented, has been met with much concern. The value of linguistic diversity has been compared with that of biodiversity, and the threats to both put on a similar moral scale. There is discomfort among some scholars, portions of the media, and various political circles as well regarding the spread of English as the medium of globalization. There are worries that English will overwhelm all. Does the planetary advance of this tongue qualify as a kind of rising tyranny, threatening to bring Anglo ways of thought and culture to every society, like an invading force? Should we accept that English may be the guilty party in high crimes of "language murder" and even "linguicide"?[9]

I will argue against these latter types of worries and indictments. But I will give much consideration to issues such as language shift and loss, inequity, and linguistic policy. The stresses and challenges posed by global English cannot be denied, belittled, or dismissed. They can, however, be exaggerated, misinterpreted, and misapplied. A great many researchers around the world have accepted the language transformation that has taken hold of science—after all, they made it happen. But acceptance doesn't obviate the need for discussion or eliminate the reality of debate. In nations where a modern scientific establishment is now being built, where English remains at an early stage in becoming part of technical training—in China, Brazil, large parts of Southeast Asia and Central America, and parts of Europe as well—the transformation defines a topic of much interest, even concern, but also and above all, ambition.

Signs of Global Science

It rises out of marshland and desert bordering the Red Sea—a gleaming new campus of glass, steel, and research 80 kilometers north of the city of Jedda, Saudi Arabia. Conceived in a vision, erected in splendor, it was built overnight—a mere five years. Its mission? To "educate scientific and technological leaders, catalyze the diversification of the Saudi economy, and address challenges of regional and global significance, thereby serving the Kingdom, the region, and the world."[10]

King Abdullah University of Science and Technology (KAUST) is sited along the reef-rimmed eastern margin of the Red Sea. Nurtured and endowed by some $20 billion of oil income (exact figure unavailable), KAUST broke ground in 2005 and first opened its doors in September 2009. It is intended to put Saudi Arabia on the map of national scientific powers, establishing a reputation that administrators hope will inspire pride among its people and aspiration among its youth. By proclamation, KAUST seeks to create a Bayt al-Hikma, a "House of Wisdom," for the twenty-first century, a direct call on the ninth-century institution that acted as a gateway for the science and philosophy of ancient Greece, Persia, and India to enter the Islamic world and help it flourish into a "golden age" of unparalleled achievement. Never mind that the original Bayt al-Hikma was in Baghdad, or that its primary contributions to world science were more in the form of translations than original discovery. It remains a powerful symbol of Islam's great era in science, which is beyond question. KAUST, meanwhile, is more of a mansion than a mere "house." Comprising both a marine sanctuary and a graduate research facility, it is fully outfitted with state-of-the-art labs; high-end equipment of every kind; sleek, digitized classrooms; and much more, all contained within elegant, greentech buildings designed to control climate, sound, and mood to the utmost degree conducive to intellectual work.

Most of all, KAUST is equipped with rising-star researchers and graduate students. These have been recruited away from institutions in over sixty nations. Within KAUST's walls, able minds from Korea, China, France, Italy, Germany, Russia, the United States, and Saudi Arabia commingle. Its first president is the dynamic Shih Choon Fong, a research engineer of the first rank who, in the early 2000s, headed the National University of Singapore, which he elevated to a scientific powerhouse with a global reputation. Under his direction, the goal for KAUST is not merely to compete with the likes of MIT and l'École Polytechnique in research but to outdo them in cosmopolitanism. This, it seems fair to say, has already been achieved. As a creation of the twenty-first century, the university is a sign of how truly international, indeed global, scientific work has become in the era after communism's collapse.

At KAUST, English is the official language. To state the obvious, an institution of such diverse personnel couldn't operate without a common means of communication. It may not be surprising that KAUST has chosen English for this purpose. Yet given the proclaimed model for this university and its location in a part of the world where Arabic has been the dominant lingua franca for well over a thousand years, the choice of English seems, perhaps, striking.

Global yet Plural

The rise of Arabic as one of the world's great lingua francas followed in the aftermath of the remarkably swift expansion of Islam as a world religion. Within a mere forty years after Muhammad's death in 632 CE, Muslim conquests had spread as far west as the shores of Tunisia and as far east as the Hindu Kush of Pakistan. By the early eighth century, Arabic had replaced dozens of tongues as the language of political and military power, and only a few decades more were needed for it to become the new language of literary expression and scientific knowledge.

Only one other language in history has spread so widely and deeply in so brief a time. In the early 1900s, English was but one of several important tongues in the sciences, below German in its total portion of published papers and roughly equal to French. By the middle 1950s, as Europe and East Asia struggled to rebuild after global war, an explosive phase of growth in scientific literature began, with papers in English now filling half the total output. This still meant that in 1960, as much as 40% of the literature continued to appear in German, French, and now Russian. But the landscape had already begun to shift. US power and wealth in the wake of World War II fueled the great engine of American "big science," driving output by US scientists ever upward. By the 1980s, English constituted as much as 70% of international scientific publication, and a decade later this stood as high as 90% for a number of fields (in chapter 3, we will discuss all this in more detail).

Many researchers have stories to tell of this change. If Ben (of my opening anecdote) were to speak on the subject, he might well reflect that despite the terrible circumstances, he came to America and entered the scientific profession at just the right time, when his English skills could do him a great deal more good than they would have a decade or two earlier. As for me, in the 1980s I wrote reports on North African geology for which

I consulted French and German sources frequently. Some of the best data in the journal literature had been written by intrepid explorer-geologists in the 1960s, capping careers begun decades before World War II. In the best of these papers, the very heat and dust of the Sahara massifs seemed to blow from the page. By 2001, when I was asked to work on a new article about Tunisia, these writings had been eclipsed. A vast new literature, all in English, now existed. This literature was of very high quality and was authored by researchers from universities in Tunisia, Algeria, the United States, and Europe, as well as from oil companies headquartered in Italy, Japan, the United States, and North Africa. A highly valuable source turned out to be the U.S. Geological Survey, then in the process of publishing its evaluation of the total petroleum resources of the globe (*World Petroleum Assessment* 2000). When I mentioned to my Tunisian colleagues the sources I had previously employed, I received indulgent, smiling replies: "Yes, they were once important. These days, though, we only use papers written sometime in the Holocene."

The rise of the use of English in science has therefore been extremely rapid, historically speaking. Its final phase occurred well within the adult lifetime of many researchers working today (including me). The reasons for this rapid rise we will explore later; they have much to do not only with American power but with geopolitical, economic, and intellectual developments in other parts of the world. It bears repeating that the larger context is English's status as a global language generally. What does this mean? It means that English holds a recognized, singular, and elevated place in many areas of international exchange. Which areas, specifically? A sampling would include politics and diplomacy; business, banking, and finance; global trade; popular music, film, and sport; tourism; airline industry; humanitarian aid; and, of course, science and technology. There are others, too (a more full list will be given in chapter 2). Yet the point is made. English has become globalized in many core areas of sociopolitical, economic, and international cultural experience.[11]

Such dominance is both tied to the superpower status of the United States and independent from it. This might seem a paradox. Instead, it may reflect a particular historical stage. US political, economic, and military power is obviously a primary force in the world and has much influence on globalization. If, for example, the military factor is largely absent in Southeast Asia, the economic and political are omnipresent,[12] while all three factors help motivate English learners in East and South Asia, Latin America, and parts of Africa. For the least-developing countries, these can be enhanced by the presence of international aid organizations such as the World Bank and International Monetary Fund, both headquartered in Washington, DC, and influenced by US priorities. American popular culture has become ubiquitous. Often adapted to produce local forms (rap music in Arabic; reality TV in Russia), the popularity of American versions still ranks very high. Recent language schools set up by the Walt Disney Company in China indicate that the trend is far from over.[13]

How, then, is the spread of English independent of US influence? There are two ways. For one, professional disciplines with an international dimension, including not only trade or tourism but most domains of business and also scholarship (even the social sciences and, increasingly, the humanities),[14] have seen growth in the use of English reach a level at which it is largely self-perpetuating. Exchanges between product managers in Thailand and Indonesia, or journal editors in Argentina and France, employ the English tongue whether they are dealing with American products and authors or not. This has continued to expand even during a period, beginning in the 2000s, when the global standing of the United States has been increasingly questioned and even interpreted to be declining.[15] Recent data from many parts of the world indicate that job candidates with better English skills can often gain salaries 30%–50% higher than other applicants, irrespective of any economic relations their country may have with the United States.[16] By 2010, the business of English-language teaching itself had become a $50-billion-per-year global industry, little of it under the control or direction of American companies. In more general terms, globalization and the Internet expanded the use of the English language far more widely and rapidly between 1995 and 2010 than did US power and economic preeminence during the previous four decades. The universe of English is no longer centered on the United States and the United Kingdom. Most communication in the language is between non-native speakers, and this will only grow. English has become something different from a "native" or "second" tongue; it is now a global means of communication whose use and advance are much greater than American influence.

Meanwhile, there is the phenomenon of new varieties of English. This language, like any other, has continued to evolve; it can't be considered in a final state. As a tongue with global reach and a colonial past, it has been adapted to many local settings. During one of my conversations with Ben, who has lived in the United States for nearly three decades, I began to notice that when we came to certain emotionally charged topics, his speech took on some interesting features: a somewhat stronger accent, certain interesting uses of words ("That first boss, he was eating the money every week . . ."), foreign vocabulary ("*kawa* OK"), dropped articles ("It was only possible because of concerted effort by my family"), nonstandard syntax ("Chairmanship of our department fell vacant due to a malpractice . . ."). I noted these changes, because they fascinated me. It was as if a layer had been peeled off, a new linguistic identity revealed. Ben had another English "self," a geographic and cultural one, which existed alongside his American one.

Over time, different peoples, different linguistic-cultural groups that use English as a second language, have added their own ingredients from their native tongue. At the spoken level, the language has become decidedly, inevitably plural. Linguists, in fact, do not speak of "world English" any longer but rather "World Englishes" or "New Englishes."[17] Distinct, nativized varieties are found in South Asia (India, Pakistan, Sri Lanka, Bangladesh), West Africa (Nigeria, Liberia, Ghana, Gambia), East Africa (Uganda, Kenya, Tanzania), southern Africa (South Africa, Zimbabwe, Namibia), Hong Kong, Singapore, and the Caribbean, all of which vary from one another in pronunciation, vocabulary, and, at times, grammar. More than mere dialects, they are less than separate tongues; *varieties* is the term used.

These world Englishes—a fact emphasized by linguists everywhere—are not deformations of some "pure" original, which, as any Brooklyn visitor to Liverpool can attest, has never existed. The question of mutual intelligibility comes up, of course: put a Scotsman, a Nigerian, and a Jamaican in the same room, start a debate, over language politics for example, and there may well be difficulties, at least temporarily. Our three speakers will struggle a bit, since they lack a final model. English, after all, has no single international body whose assigned duty it is to monitor and standardize the language for all users—to search out and destroy, covertly if necessary, all deviations and unwanted growths from the Anglo-American standard (often known as Standard English, or SE)—an impossible task in any case, as the French long ago found out with regard to Canada and

West Africa. In formal writing, much of this variety disappears; there is indeed something approaching a global standard when it comes to written discourse at this level. Yet even here things are changing, as we will see later on.

Matters Political and Educational

At a recent conference on future energy technologies, I pulled aside one of the presenters to ask a question. Elyana Meyer, a plant pathologist employed by a large Brazilian firm, had delivered a captivating talk about the use of natural selection to develop new varieties of sugarcane for higher-yield production of ethanol. Dr. Meyer read a paper—in English—describing a process by which seedlings from northern Brazil were transplanted to a southern state, where they competed against commercial varieties grown there. After three years, the most robust plants were preserved and studied, and their clones transferred to the lab. Several more iterations will be employed to reduce the total number of candidates to two or three varieties, from a starting pool of over ten thousand. The entire process will require a decade or more, coached from the sidelines by Darwin himself.

I was eager to talk with Dr. Meyer because of a book I was then writing on global energy; I wanted to inquire about the history of the project and any new developments. A bit stiff and imperial in person, she looked away as I spoke. I softened my tone, complimented her on an interesting paper, and began to pose my queries once more. "I am sorry," she said suddenly. "My English speaking is not good. Please send e-mail." She handed me her card, rotated on a heel, and accelerated away, perhaps to the restroom. I realized my error. When I returned home and sent an e-mail repeating my compliments and questions, I received a very different reply: "I wish to apologize for the rudeness. You are kind to say nice things about the presentation. It was not difficult for me to write, but talking is a challenge. Here are answers to your questions. . . . Thank you."

Issues of language and science reach well beyond the academic realm. A great deal is at stake in the globalism of English, and not only because of what it might mean for Britain or America or for French or Chinese. At every level, it involves relations among people, institutions, and therefore nations.

My brief and ultimately successful and valuable contact with Elyana

Meyer had a distinct geopolitical dimension. It can be described as involving every nation's drive to improve its energy security, the desire of many nations to lower their oil consumption and replace it with renewable resources such as biofuels, and Brazil's key role as a major supplier and exporter of one such fuel, ethanol. But even here, we are merely resting at the surface. We need to delve into how globalization and economic development in the post-Cold War world have affected a great variety of nations, making energy a central concern. China, for example, since 1990 has pulled hundreds of millions of its people out of poverty, built fully modern cities, and spread its aggressive demand for raw materials across continents. India, too, though following a different path in some respects, is generally on trend to achieve something quite similar. In their drive to collapse centuries of advancement into decades, both nations rely primarily on the energy resource they have in greatest abundance—coal—and on oil imported from the Middle East and West Africa. Although helping to build industries and provide power on a massive scale, coal has degraded many streams, lakes, and rivers; its pollution of the air in urban areas has led to a high incidence of respiratory disease and, in China's case, a stained global image. Motorizing their citizenry, meanwhile, has led China and India to surging oil imports, which by 2012 had reached over 60% of their overall consumption. Both nations want to improve this state of affairs, and Brazil may be able to help. With the largest economy in Latin America, it has become both a major new oil state as well as a producer of ethanol. Moreover, it is technologically advanced and has enormous potential for agriculture—in 2012 it was using a mere 13% of its total arable land. No other country comes close to having so much room for expansion in food and fuel production.[18]

Elyana Meyer's research—like that of thousands of other scientists and engineers—is a direct result of these realities, as is her ability to communicate with the wider world. This ability may be limited; she has been caught between generations, not having learned English very well in school but forced to acquire it as an adult for career demands. She is unable to be fully at home in this second language as yet. But this is precisely the point. Geopolitical changes help drive language changes, to which successful scientists respond.

Nearly every modernizing nation, including the so-called BRICs (Brazil, Russia, India, and China), has identified science, technology, engineering, and mathematics (STEM) as the generators of future prosperity and the best way to handle problems related to disease, food, water, and energy. Strategic reliance on STEM has been embodied in official plans, such as China's National Science and Technology Development Plan (2006), and in expert bodies, such as India's Science Advisory Council.[19] China has backed this idea further with real support, growing government investment in R&D no less than 20% every year for the past decade.[20] R&D spending as a percentage of overall gross domestic product has risen in all the BRIC nations and throughout the developing world, a fact discussed in more detail in chapter 3. What all this means for language can be summed up in two words, *education* and *training*—education, that is, of students in STEM and, most often, English as well; training of teachers, also in these same two areas. To gain some idea of how fundamentally education and training are being viewed by rising world powers, consider these ultimate goals of Brazil's educational system, written into its National Education Guidelines and Framework Law of 1996:

> National education, inspired by the principles of freedom . . . has the purpose of: a) understanding individual rights and responsibilities, as well as those of citizens, the State, and other community groups; b) respecting the dignity and fundamental freedoms of human beings; c) preparing individuals and the society to master scientific and technological resources . . . for common welfare; d) protecting . . . the cultural heritage; and e) condemning any unequal treatment resulting from philosophical, political or religious beliefs.[21]

Both China and India, meanwhile, are intent on building top-caliber research universities. They have seen the rise in that regard of Japan and South Korea, whose college student populations increased from under 10% to nearly 50% in a single generation after the 1960s, and whose STEM capabilities rose to world-class levels. China has launched a massive effort at catching up: in the decade after 1998, when President Jiang Zemin launched a program (Project 985) to build up the country's higher education system, the number of its universities doubled to over two thousand and its enrolled students increased fivefold, to more than 5 million. On the basis of STEM departments, the government identified nine universities to make up its "C9 League" of top-tier institutions. India seeks to upgrade its fifteen Institutes of Technology, the "pearls" in its own system, which have yet to halt the flow of students overseas for scientific training, especially in the United States.[22] Further investment in these institutions is seen as critical to India's long-term future. Similar ambitions have been publicized for Singapore, Thailand, Vietnam, and Taiwan. But the phenomenon is global. Achieving mass higher education has become a goal of every nation that can afford it (and some that can't), focused above all on producing scientists and engineers.[23] "The great brain race," as one recent author has dubbed it, is well under way, and is being run nowhere more intently than in the fields of science.[24]

All these efforts include the learning of English. Study of the language became required by the mid-2000s or earlier in major Asian nations, starting in primary or secondary school. In China, as in Korea and Japan, it is one of the core subjects, along with mathematics and Chinese itself, included on university entrance exams. In both developed and developing nations, a major trend is to employ English as the medium of instruction, especially at the university level. This trend has been under way in China for science courses since 2005, with a growing number of institutions now requiring their faculty to publish one or more papers in English as a prerequisite for tenure or other advancement. My anecdote about Dr. Meyer indicates that this initiative is happening in the corporate sector as well.

Three other developments to consider in this context: (1) the growing international mobility of students and researchers, already discussed; (2) the building of branch campuses by Western universities in foreign countries; and (3) the increase in for-profit education providers, including private academies and "cram schools," which make English a core subject. In blunt summary: more students studying in anglophone nations; more English-speaking campuses worldwide; and more local centers where English is being taught to those in the sciences and engineering.

In all these phenomena, we see that the very economic and intellectual ambitions of the world are creating a formidable pressure for English to expand. It would thus require a profound and disrupting change to the global scene for any other language to replace English, not least in the international scientific arena. Such change is always possible, though; it has happened before, even in Europe. French, for example, had overwhelming prestige in the eighteenth century, but this was largely gone a hundred years later, due to the wars, social upheavals, and industrialism of the 1800s. French remained an important regional tongue, even well after the defeat of Napoleon and the rise of British and Prussian power—no less a scientist than Charles Darwin, sent to Paris to learn French for curing a childhood stammer, felt that the language was still as necessary for his work. Patterns of language use, embedded in a variety of institutions and social practices, not to mention living generations, do not vanish or alter overnight but require time and transformation. By nearly any measure, English will be a global tongue for decades to come; indications certainly suggest it has not yet matured in this capacity.

For many countries, not least those that suffered the effects of colonialism, demanding that all schoolchildren learn English can raise debate or controversy. At the same time, however, these countries are beginning to realize that nations in East Asia, after embracing such a linguistic policy, have not in the least abandoned their native tongues, fled domestic traditions, and mutated into obese Americans. As these types of controversy fade, the pressing issue won't be the extent of English but rather the quality of its expansion. Where will all the thousands of new English teachers come from? Who will they be? Will they all need to be native speakers, and if so, how competent will they be as teachers? If not, how well will they be trained? And for students who go abroad, what level of English will they return with, if indeed they do return? How might their English connect or clash with that taught and learned domestically?

Geopolitics, therefore, offers its own forecast. As the developing world advances, it will become a greater and more complex user of English. Almost certainly, Asia will be the global center of such use in the twenty-first century. This won't happen without challenges. Elyana Meyer, with her strong writing ability and self-conscious speaking, demonstrates that the process, whether in Asia or Latin America, will not be easily achieved. She also suggests, however, that it can be done.

Final First Words

During the first half of 2010, I helped design a scientific writing program for a major biotechnology firm that employed a sizable number of Chinese researchers. These researchers, I was told, plus others from South

Asia and Latin America, did excellent work. But the papers they wrote for publication and for government regulatory agencies often arrived, as one manager delicately put it, in "English on crutches," needing a good deal of steadying help. Other company scientists had to be brought in to rewrite portions of these papers, taking away from their own research work. Sending material to editing agencies proved expensive and time consuming. Ghostwriting was not an option either; though it had been employed in the industry, it had led to several scandals and much trouble. No single set of techniques existed that could solve the problem.

The viewpoint I brought to the table was that much could be achieved with attitude, new ideas about science and the written word. To improve the writing of nonnative speakers, it would help to understand that communication is not separate from research, and that better communication makes for better science and increased career opportunities, confidence, and achievement. Learning to write well in a foreign tongue takes much patience, forgiveness, and hard work, a good deal of it attained via disciplined imitation of well-written texts. But the rewards can be enormous. In fact, Liang Chin, originally from Shanghai and well published in English, made me see that these rewards can be greater than I had suspected. Sitting with me one day during a workshop coffee break, Liang said,

> Since I first came here and started working, I write many papers, all in English. Then I must e-mail and lecture and seminar in English, too. It is very hard for me, very hard. But I know I must do it. I went to English school at night, and the company paid for it. It means they want me to stay. My wife also says I must do it [*laughs*]. I did course for one year. It can raise my English standard. Four years after, writing English is still hard. But not like before. Sometimes I ask friend to check my article. This is very helpful. Also I used to feel jealous of colleagues in China. They could write papers in Chinese! But I know my feeling was foolish. They cannot leave. They are birds in this great cage. I can go work in many, many countries.

Liang Chin will never be a perfect speaker or writer of English. He will always have an accent and will always, I expect, make grammatical errors when measured by Anglo-American standards. None of this, however, matters very much. He is entirely fluent in his ability to communicate, above all to present and publish his scientific work to English-speaking colleagues everywhere. His lack of perfection doesn't render him a second-class citizen in the lands of scientific English; on the contrary, his output in international (English-only) journals exceeds the majority of those in his field, making him more successful than most native speakers.

In short, the rise of scientific English defines a profound development affecting the lives and careers of millions. For some, like Ben in our opening story, it has created life-giving opportunity. For others, like Elyana Meyer, it is part of the given order and acts to both enrich and complicate their work. And for still others, like Liang Chin, it has brought frustration, isolation, but also, in the end, triumph. Learning English for any of these people was not an act of submission but one of expanded capability.

No area of human knowledge has desired a world discourse more than science. The symbols of mathematics embody this goal, though they came from one country, India, and are quite recent in their standardized use. But mathematics is a language only in a metaphoric sense—it is not a form of speech; we cannot use it for the great majority of communicational acts we perform each day. During the Scientific Revolution of the seventeenth century, ideas were advanced for a perfect, constructed language based on mathematical principles. John Wilkins, René Descartes, and Marin Mersenne all proposed such utopic tongues, not only to advance the study of nature, but even more to cure the ills and evils arising from misunderstandings among people and nations. A "universal character," as Wilkins called it, would be entirely neutral in that it would not favor any group, and it would link the progress of science to the spread of peace and a shared, sympathetic mind.

History suggests that the ambition has been both noble and futile. A perfect tongue to unify humanity, whether inspired by the Bible or modern linguistics, is a fable. In more recent times, we have had many attempts, from Solresol to Esperanto, and all have failed. With no attachment to any nation, any culture, any living people, with no unique history, origin, or status, a synthetic tongue lacks the very content that makes a language fully human. People, scientists included, have not been drawn to such a language, because of its very artificiality and lack of worldly connections.

Researchers have long conceived the value of a unifying tongue. They now have one. Or rather, to indulge in a bit more precision, they have what may soon become the closest thing to such a tongue humanity has ever been able to provide. It is very much a living language, evolving and adapting. Linguists will insist—rightly, as Germans, Russians, the French, and others will testify—that the choice of English as the language of science has absolutely nothing to do with any of its inherent qualities. There is no wondrous fit between English and things scientific, just as there is no higher essence that links German and philosophy or French and diplomacy. Any other major language could probably serve as well as English in STEM.

Some language scholars have objected to the global role of English, specifically in science, and the troubles it can bring.[25] What, then, if Chinese were to attain this role? Or Spanish, or Arabic? Obviously, the basic challenges to nonnative speakers would be the same. As we have already noted, a global tongue, no matter what specific tongue it may be, brings with it certain realities. English, however, is said to be imposed by the forces of globalization, led by America. Yet the global reach and expanding power of China would surely come in for criticism, were Chinese to become the tongue of science. Spain's own imperial past and the political and cultural controversies surrounding Islam would no doubt rise up as issues, were Spanish or Arabic to gain such status. Other international tongues, such as Russian and German, would certainly bring their own difficulties. Languages are too easily blamed for the perceived faults of their home countries, an unfortunate confusion. To be fair, most scholars who have objected to English as the language of science would prefer a multilingual landscape, with perhaps three or more dominant tongues. But this would defeat the enormous benefits of a single lingua franca in the scientific domain, where knowledge aims at universal acceptance and thus a global audience, and where scientists themselves strongly desire such a globally shared language.

Historical factors matter. That English has been the tongue of the greatest colonial power, of the homeland of the Scientific and Industrial Revolutions and of the greatest postcolonial economic, military, and scientific power, has some relevance. Yet more recently, these aspects no longer drive all the choices that are spreading the dominance of English in science. No imperial policy is at work to impel scientists to take up this one language. We are in a new period, where new historical patterns are at work, not least of which are the globalization of science itself and of knowledge generally, propelled by an extraordinarily powerful new medium of communication and contact perhaps no less foundational for the future than writing has been for the past.

It is the job of this book, then, to provide a first-order determination (as we scientists say) of whether a global tongue is truly a good thing for science, and why. It cannot do so, however, without first examining the landscape of global English. Readers will not be surprised to learn that this landscape defines a subject about which a great deal has been observed, debated, rejected, and revised, with much new thinking and fruitful insight. It is to this larger context, in which the language of science abides, that we now turn.

(2)

Global English

Realities, Geopolitics, Issues

The allies of Rome, who had borne the principal burden of . . . extending the empire, demanded to be admitted to all the privileges of Roman citizens.

ADAM SMITH, *Wealth of Nations*

Ethiopia is one of the most ancient of African nations. Briefly colonized by Italy during the fascist era, it has since been independent, but very poor. Most families in 2012 earned less than nine hundred dollars. After the last emperor, Haile Selassie, was deposed in 1974, the country was torn by violent coups, famine, and massive refugee problems. Only since the mid-1990s has it benefited from a functioning constitution for its 90 million people, half of whom are fifteen years of age or younger and suffer from hunger. Less than 50% of Ethiopians are literate, but this percentage is increasing; 5.5% of the gross domestic product is now being spent on education, the same as in many developed nations. Dozens of ethnic groups speak more than sixty indigenous languages, with Amharic the official national tongue. Written in a unique script, Amharic is second only to Arabic as the most-spoken Semitic language.

Digum Elementary School sits by the side of a dirt road in the Tigray region of northern Ethiopia, about five hundred miles (800 km) from the capital, Addis Ababa. In its grade 1 class are forty-four boys and girls aged five to seven years. They come from surrounding villages, and most are barefoot. Few have ever seen a TV, a computer, or an iPhone. Digum has no piped water and little electricity, none in private homes. Yet in this overcrowded classroom, every child knows the English alphabet. English, in fact, has been chosen as the main foreign language for study in all of Ethiopia's schools.

A British journalist who works for a major newspaper comes to visit. He is staying at a campsite near the town (there is no hotel), communicating with his editor via satellite phone. Touring the classrooms, observing the students reciting and singing in English, he takes a moment to ask a boy in a grade 8 class a question: why does he want to learn English? The boy says, "It is the language of the world, and I want to know the world." Then, another boy asks the visitor, what native language does he speak? Is he Italian, by chance?

The visitor from England is stunned. He realizes that even in this remote corner of one of the world's poorest countries, knowing English is no longer attached to any nationality, including his. He writes, with a bit of emotion, "*We are losing ownership of international English . . . [it] isn't really ours any more.*"[1]

How Global, Really?

David Crystal and David Graddol are two of the best-known and most widely cited experts on the history and present circumstances of the English language. Interestingly, both authors wrote volumes in the mid-1990s that sought to diagnose the outlook for English, and then updated their forecasts in the early 2000s, less than a decade later. Crystal's first version of *English as a Global Language* (1997) begins this way: "In 1950, any notion of English as a true world language was but a dim, shadowy, theoretical possibility. . . . Fifty years on, and World English exists as a political and cultural reality. . . . [But] have matters developed to the point where the rise of English as a world language is unstoppable?" He perceived at this date that his subject was only "the leading candidate" for a true global tongue. A mere six years later, however, when the second edition of his book was published, Crystal felt fully able, even obliged, to answer his own question: "[English's] growth has become so great that there is nothing likely to stop its continued spread as a global lingua franca, at least in the foreseeable future."[2]

Graddol's own 1997 book, *The Future of English*, took a different tack. It predicted that no single tongue would dominate the twenty-first century, but that five or six languages would hold court over international communication—Spanish, Arabic, French, English, Russian, and Chinese. Such a forecast reflected the early years after the fall of communism, when the world seemed open to a multipolar future and to growing linguistic diversity at a global level. With *English Next*, however, published in 2006, Graddol found himself forced to offer an amended message. English had become so prevalent, so dominant, so unchallenged that the ability to use it for global communication would soon be a "basic skill," like using a computer keyboard. By becoming so ordinary and expected, proficiency in English will lose special status, Graddol said, leaving native speakers in their own shadow. Lacking skill in it, however, would be crippling.[3]

Both Crystal and Graddol have been, in their own ways, startled by the rapidity with which English has spread across the world. They are careful to stress that the phenomenon is not at all universal—indeed, the greater part of humanity does not speak or study English—and it will more than likely never be so. It is easy, they say, to overestimate the reach of the language—except, perhaps, in certain professional domains. How global *is* English, then? How many actual speakers does it have worldwide? *Global*, of course, implies not only large numbers of users spread over continents but also many landscapes of competence. If English is the most extensive of international tongues today, by how much does it exceed the likes of Spanish, Arabic, French, or Russian?

These are difficult questions. No one can precisely answer them. To do so, we need a register of all the living languages on earth and who speaks them, along with the degree of multilingualism among each linguistic community. We might assume that linguists have fully explored the planet in terms of human speech. But they haven't, and not because of laziness or incompetence. Besides intractable problems of definition—where does a "language" leave off and a "dialect" or "creole"* begin?—there is the geographic and cultural challenge of locating thousands of speech com-

*A creole is defined as a natural, native language that represents the deep mixing of two or more parent tongues and is spoken as the primary language of a group. Most creoles have passed through the stage of being a "pidgin," a more primitive, grammatically reduced mixture of other tongues.

munities having only a few hundred members or less that may exist as a subset within one or more larger communities. How to do this, moreover, when languages are evolving, with some even becoming extinct each year, while others are being born? It turns out, therefore, that in some of the most linguistically diverse areas of the world, where native tongues are especially abundant, only crude or rudimentary surveys have been made.

The closest thing to a linguistic register for planet earth, in fact, is assembled in *Ethnologue*, a publication of a US faith-based organization.[4] The *Ethnologue* staff has worked heroically to compile its database, striving to be as complete as possible but relying heavily on census data from many countries. It turns out (we can pretend to be surprised) that not every nation keeps a scrupulous log of which languages its citizens speak. The United Nations has long recommended that countries conduct a complete census once a decade. But in many parts of the world this hasn't been possible, because of war, internal conflict, disputed borders, poverty, and so on. Data that do exist can easily be outdated, incomplete, or unreliable.[5]

In short, we can do little more than make educated estimates of what people speak worldwide. Take India, for example, one of the most multilingual nations in the world. Here, a census is indeed taken every decade, but the results are anything but clear and consensual. In 2001, the census documented as many as 6,661 named "mother tongues," yet linguists determine that many of them (thousands, in fact) are actually synonyms for the same languages or else dialects of them. As for English, figures given for the number of Indians who use the language competently vary enormously, from at least 55 million to as many as 350 million.[6] Given such uncertainty, what are the best, most acceptable estimates for English at present? Those who have spent much time and effort in the fields and trenches of global English (Crystal and Graddol included) suggest that the following points are likely to be true:[7]

- English is a native language for about 360 million to 380 million people, and a *second* (added) language for an equal number. As a *foreign* language, it is used to a variable extent by about 800 million to 850 million. Most of these latter speakers are in Europe, Asia, and Africa.
- By 2010, around 1.5 billion to 1.6 billion people, roughly one quarter of the world's population, were therefore using English

with more than rudimentary skill. The language has been declared an official tongue or given special status in at least seventy-five countries that extend well beyond the British colonial empire.

- Also in 2010, another 400 million to 500 million children aged fifteen years or younger were studying English in school, about one-fifth of the world's child population. English is the primary foreign language taught by schools in over one hundred nations, including China, Russia, Brazil, Pakistan, Nigeria, Bangladesh, Indonesia, Egypt, and Iran.
- The number of English learners (all ages) is projected to peak at around 2 billion before 2020. Declining birthrates will reduce these numbers thereafter.
- The global spread of English has a generational component: young people (under the age of thirty) have taken up the language, in school and professional life, at an especially high rate.

These statistics are both welcome and daunting. They are welcome in that they help paint an early portrait of global English; they are daunting in the type of portrait this seems to be. English has gained speakers and learners by the tens and hundreds of millions on every continent (it is the lingua franca on Antarctica, too), in every one of the world's major nations, and in countries that are direct rivals to, or enemies of, the United States. Such numbers can be formidable even for native speakers. What it means, after all, is that today, far more interactions in English take place between multilingual individuals in non-Western countries. In East Asian nations including Japan, Korea, Taiwan, and increasingly China, it is taught to all secondary-school students in both public and private institutions, regardless of background or income level. In many developing nations, English is viewed as a required pathway to professional opportunity, economic advancement, and popular culture. Or, as noted by Yun-Kyung Cha and Seung-Hwan Ham, who have spent two decades studying the diffusion of English in school curricula across the world, "A certain high level of ability to communicate in English seems to be becoming in many countries a new kind of basic literacy that no longer conveys Western ideological connotations, just as computer or information literacy is considered a basic requirement for today's world citizens."[8] As the visitor to Digum School realizes, the language no longer belongs to England or America.

Originally a colonial tongue in many parts of the globe, it has become a means for people to try and move beyond the effects of colonialism.

English in Global Education

Another way to regard the spread of English is to consider the idea of "soft power." This has been done by other writers (David Crystal, for one) and is particularly relevant here. *Soft power*, a term coined by Joseph Nye, an international relations scholar and former US assistant secretary of defense, refers to the idea that a country can gain influence not only by force but by attraction, through such things as effective policies, good institutions, wealth, and ideology.[9] But language is obviously another type of attractor, one that supersedes any national boundary. A large part of the reality for English today comes from the desire of other nations and peoples to imitate and draw from its success. This, too, is what we learn when we hear the boy at Digum School say, "It is the language of the world, and I want to know the world." Can we measure this form of attracting power in some way?

One approach, suggested by our list of likely facts above, is to look at global education. It was noted that at least one-fifth of the world's children now study English, and that this portion will increase, since by 2010 English had become the main foreign language taught in primary and secondary school for 130 or more nations. Cha and Ham, whose work on the place of English in global school curricula was mentioned earlier, have provided a lucid and unmatched set of data on this topic over an extended time period, from 1850 to 2005.[10] Their information, covering many dozens of countries, shows that before the twentieth century, English was barely taught as a main foreign language in primary and secondary schools of non-anglophone countries. Among five world languages—German, French, Spanish, English, and Russian—German and French remained dominant in classrooms around the world until as late as the 1920s, '30s, and early '40s. After World War II, these tongues were rapidly and progressively replaced, so that by 2005, English was the primary foreign language taught in approximately 70% of primary schools and 80% of secondary schools in 157 countries and territories. Regions where this proportion was 85% or higher in secondary school included the Middle East/North Africa, Asia, Latin America, and western Europe. Most of the nations and territories involved were never part of Britain's colonial empire or within the United States' regional sphere of influence.

There are more precise numbers, however, for a different and no less revealing group of learners: international students. According to the most recent available data collected by the Institute of International Education, the Organisation for Economic Co-operation and Developments's *Education at a Glance*, and the U.S. Council of Graduate Schools, the following statements about this group can be made:[11]

- The total number of international students rose from 2 million in 2000 to nearly 4 million in 2011—a far more rapid rate of growth than in the previous twenty-five years, when their numbers increased from 1 million to 2 million. About 44% of these students in 2009–10 were attending universities in five anglophone countries: the United States, the United Kingdom, Australia, Canada, and New Zealand.
- The total proportion of students using English grows to 50% if we add countries where English is now generally used in college instruction: South Africa, Singapore, Netherlands, Sweden, Denmark, and Finland. It rises higher still if we include students who use English in some of their coursework, as mandated by selected programs in France, Germany, Belgium, Switzerland, the Czech Republic, Norway, Iceland, Poland, Korea, and Japan.
- The United States, as the top destination for foreign students, has been gaining in total numbers but losing in global share, falling from 23% in 2000 to 18% in 2009. This has happened even though popular rankings of world universities continue to favor US institutions by a wide margin (up to 80% of the top twenty-five institutions). America's percentage loss, however, has been more than made up by gains in Australia, Canada, and New Zealand, and by growth in English-language programs offered by non-anglophone nations.
- The percentage of applications to US graduate schools from foreign students rose an average of 8.7% per year from 2005 to 2012, with the greatest increases seen in those from China (59% growth from 2010 to 2012) and the Middle East plus Turkey (12%). Most frequently chosen fields of study were engineering (26% of all

foreign graduate students at US universities in 2012), physical and earth sciences (20%), business (17%), and life sciences (13%).

- The most popular fields of study chosen by undergraduate foreign university students in the United Kingdom, Canada, and Australia include, in rank order, business and management, engineering and technology, economics, health sciences, physical and biological sciences, and mathematics. Nearly half of all foreign students in the United States, Australia, and the United Kingdom come to study science, engineering, and health-related fields.
- The number of sending nations has exploded since the Cold War, and now includes Eastern Europe, Central Asia, Southeast Asia, Russia, and China. By 2009, more than half the international students (52%) were from Asia—above all China, India, and Korea. That year, 567,982 Chinese were studying abroad, more than double the next nation, India (211,038), with 43% headed to the United States (124,225), Australia (70,357), and the United Kingdom (47,033). Before 2007, the majority of Chinese studying in anglophone countries were graduate students. This percentage has rapidly shifted toward undergraduate and even high school students.
- English-speaking countries have been the big seller's market for foreign students, but may face more competition before long. Many former sending countries plan to draw more international students, with Thailand, Malaysia, Singapore, Turkey, Brazil, and China among them. Moreover, places such as Japan, Korea, Taiwan, and Hong Kong, with very low birthrates but well-developed higher education, may well see the need to recruit good students (future professionals) to help keep their economies strong. In most cases, the market for students will be forced to obey the market for knowledge. And as we've noted, in engineering and the sciences this means knowledge in English, hence instruction in this language.

In other words, international students vote with their feet. To a certain degree, given their concern about pursuing studies that will make them more valuable to the job market (improving, as economists say, their human capital) and help them launch a career, these internationals are at the forefront of globalization. A great many of them, particularly those who remain in anglophone countries for several years or more, will become good to excellent in their command of English. Some, of course, will not; but this doesn't mean they will cease employing the language in some capacity.

But the focus on foreign students in anglophone countries comprises only one part of a larger trend. Well beyond the reach of Anglo-America, English is already becoming the tongue of international instruction at the university level, and above all in business and the sciences. One does not have to be a university administrator to see that over time, this could open up and even democratize the global system of higher education, such that English-competent students and academics from anywhere would be able to study and teach anywhere else. The trend appears to be strong, and growing.

Such is apparent from many course catalogs that can be found online. For instance, in 2009 Nebrija Universidad in Madrid, a small institution with 2,800 students, listed more than fifty subjects taught in English, including general linguistics, international marketing, and communications theory. That same year, the University of Helsinki, one of the top research universities in Europe, whose student body exceeds 35,000 with an additional 30,000 adult learners, offered classes in English in all the natural sciences, as well as in the faculties of law, medicine (human and veterinary), the arts, and even theology. Within two years, all these offerings at both institutions had expanded further.[12] For nearly a decade, Japan's Tohoku University (founded 1907) has provided doctoral programs for Japanese and international students in nanoscience, aerospace engineering, bioengineering and robotics, and more than a dozen other fields, all taught in English. In 2009, moreover, the university launched a major reorientation plan to build itself into a world-class institution: "As a Creator of knowledge, we will strive to produce future leaders with a strong liberal arts background, specialized expertise, and an international outlook." To internationalize its educational offerings, Tohoku plans to expand online learning, create overseas internships, emphasize cross-cultural communication, and, at a practical level, "increase the number of hours of English lectures."[13] Beginning in 2001, meanwhile, the Chinese Ministry of Education began to require that textbooks in English be used in pretty much *all* university-level teaching for a number of major areas—information technology, biology, finance, law. By 2007, Chinese universities with English-taught programs at both the undergraduate and the graduate levels ran deep into the dozens and included many of the most prestigious institutions.[14]

A Bit of History

How did all this happen? Scholars of the English language clearly indicate that it was centuries in the making. A foundation was set with the geographic distribution of the British Empire, which was not concentrated in one or two portions of the globe but spanned every continent and many island spaces in between. The power and influence of Britain's control over the seas, and consequently world trade, in the eighteenth and nineteenth centuries were pivotal. Over time, so was the regional impact of its native-speaking territories: Australia, New Zealand, South and West Africa, Canada, and, of course, the United States. The rapid spread of English beyond these places faced real hurdles early on—England's language policy for its colonies before the 1820s was to promote education in local languages to maintain separation and dependence. Later, English was still restricted, mainly to higher levels of the colonial administration, translators and other mediators, and important merchants. But as the language of power and commerce, it was soon appropriated by others, particularly those who had an interest in technical matters. England, lest we forget, was home to both the Scientific and the Industrial Revolutions.

After British imperial power peaked around 1900, the United States began to become the force behind the spread of English. America's launching of the second Industrial Revolution (1880–1914) and the enormous wealth it generated led the United States to finally become a true world power, joining England and France as major victors in World War I. During this same time period, the United States was the source of an entire panoply of inventions and innovations that changed the nature of daily life in the West: the telephone, the phonograph, electricity, electric lights, the airplane, the motion picture and automobile industries, jazz, and mass culture generally.

On the political and economic side, the United States became a considerable force owing to its victories in World War II, its success in rejuvenating Europe through the Marshall Plan, and its grounding a new global economy via the Bretton Woods Agreement, as well as its widespread military bases and reconstructive efforts in East Asia. American popular culture, such as Hollywood-produced films, rock music, and radio and television, spread rapidly in this period, when the Cold War deepened. Certainly, no other nation, especially the Soviet Union, could compete with the United States on this cultural level. Even during the troubled 1970s, America's economy remained the largest and most dynamic in the world while its leadership in politics, energy, and science rose in stature. When the Cold War ended and economic globalization took over, all of this accelerated. With the collapse of communism, America's triumph seemed secure, and its status as a nation of unparalleled opportunity and exciting youth culture confirmed.

Fields of Dominance

A global language casts into doubt certain traditional linguistic distinctions. Consider "first," "second," and "foreign" languages—those that are learned, respectively, from birth, after the first language, and as a tongue not native to one's country. Such distinctions are tied to a monolingual perspective from the start, since in many linguistic groups children learn more than one tongue at home; that is, they have at least two first or "mother" languages. Meanwhile, in former British colonies, such as Nigeria or Singapore, English can be learned in the home, at school, in the street, or in other settings. It can thus be classified as all three of the designations given above while also being an "official" tongue of the country. Thus, calling it a foreign language makes little sense; yet, since it is learned in school by a large number of people or at home along with an indigenous tongue, it can't be easily called a second language either. The difference between "first," "second," and "foreign" becomes more a matter of context, function, and use. Just as a business professional or scientist learned English at school and then comes to use it daily for work, travel, and socializing, so a street vendor may know two or three native languages and have picked up a ready command of broken English to deal with tourists and guides.

A global tongue increases multilingualism. English is now often termed an added language. Even before the year 2000, the following type of situation was observed in this connection:

> A Korean manufacturer in an Athens hotel meeting the Brazilian buyer of a Swiss conglomerate will not only negotiate but order dinner from his room service in English. There may not be a single native English speaker in the hotel, but all non-locals staying there communicate with each other in English. . . . In business, sport, politics, science and many other fields, a knowledge of English has become not a matter of prestige but of necessity. Also: the level at which this occurs is moving ever downwards.[15]

A country that determines it will not teach English to its people is one that seeks defiant isolation. Although this may be good for national self-image, as a policy for interacting with the larger world, not least in the sciences, it is likely to be disastrous. Problems caused by isolation from English become even more evident when we list the domains in which this language now commands a dominant or fairly dominant position in a global context. These fields include the following:

Advertising
Biomedicine
Broadcasting (radio and television)
Business
Computer industry
Computer/video games
Diplomacy
Economics
Energy industry
Engineering
Environmentalism
Film industry
Global anti-terror operations
Global communications
Global fashion industry
Global finance and financial institutions (World Bank, IMF, Asian Development Bank, etc.)
Global/international marketing
Global/international trade
Global jihad (between speakers of Arabic and other languages)
Global maritime industry
Global publishing
Global rescue (natural disasters)
Global security ("Policespeak")
Global sports
Humanities
Human rights
International aid
International banking
International education
International law

International relations (field of study)
IT services, consulting
Natural sciences
Nuclear power industry
Popular music
Social sciences
Travel and tourism

Use of English does not govern equally in all these areas, of course. For some, such as air traffic control, it is a nearly universal application; in others, such as publication in the humanities, it varies greatly, depending on the field. Much trade still takes place in other international tongues—Spanish for regional commerce in Latin America, Arabic in the Middle East and North Africa. Nations including France, India, and Japan have their own successful film and fashion industries of international fame, to say the least. Yet in the global context, English tends to rule far more than any other language. A company doing business in Denmark and Brazil (such as the wind power firm Vestas), needing to negotiate with clients in both nations, will employ English by necessity. The World Cup and the Olympics, the world's two greatest sporting events, are broadcast globally in English, with participants often giving interviews in this language. And no film culture can truly compete with the global reach of a *Star Wars* or *Avatar*. Moreover, there are many specific industries that deserve mention for their heavy use of English—automotive manufacture, construction, pharmaceuticals, mining, petrochemicals, agribusiness, and so on.

As things stand today and for the foreseeable future, no individual, group, or nation can hope to engage any of these areas at the highest level without some proficiency in English. Even the European Union (EU) has struggled mightily with this fact and finally accepted it, despite its carefully fashioned policy for the multilingual community known as "mother tongue plus two" (every citizen to be trilingual to some degree). In theory and by official decree, English has no higher status in the EU than any of its twenty-two other official languages. In practice, professionals recognize the potent and obvious advantages associated with this tongue, which is taught as the first foreign language in the secondary schools of *every single* EU *country* outside Britain and Ireland.† Nearly every major newspaper and magazine with hopes of an international readership has an English-

†According to data published by Eurostat, the percentage of all pupils in the twenty-seven nations of the EU studying English in upper secondary school rose from 82.5%

language edition.[16] Gather diplomats, docents, or doctors from Germany, France, and Spain in a single room, and it will be English with which they choose to inform or insult one another. That such is the case in the region where modern linguistic identity was born tells us a great deal.

Geopolitical Factors

We are left, then, with the question of which events combined to propel English into its global status. For it has really taken just a single generation, since about 1980, for this to happen. Of course, the stage had been set by some of the most profound occurrences of the modern era—the Scientific and Industrial Revolutions; the building of the British empire; two world wars; the consequent rise of America to superpower status. If Britain dominated the imperial nineteenth century through its powerful and diffident aristocracy, America took charge over much of the twentieth with its ideology of democracy and egalitarianism. Starting in the 1980s and early '90s, however, the world underwent a new series of geopolitical shudders that ended up granting force and favor to the English language. What were they?

First among these events were the collapse of the Soviet Union (1991) and the creation of the EU (1993, in its current guise). These happenings led on the one had to the building of a supranational entity deeply allied with the United States, and on the other to the destruction of the great adversary of English-speaking America and its global influence—a dissolution that also produced dozens of new nations across Eastern Europe, the Caucasus, and Central Asia. The end of the communist era in the Eastern Bloc (1991), in fact, soon brought many new countries into the EU, seeking integration with the West. Communism, as an anti-Western, anti-capitalist, and therefore to some degree anti-English ideology, also died a summary death in China and Vietnam at about this time. It was replaced in those countries with a massive drive for economic development and thus cooperation with Western nations, not least the United States, as well as with Westernized Asian neighbors, such as Japan. The end of apartheid in South Africa and the installment of full democracy in

---

in 2004 to 94.6% in 2009. See http://epp.eurostat.ec.europa.eu/portal/page/portal/product_details/dataset?p_product_code=TPS00057.

South Korea, a country that soon joined the Organisation for Economic Co-operation and Developments (1996), or "club of rich nations," were two other developments of major importance.

Events of a darker hue transpired as well. The wars and genocide that accompanied the breakup of Yugoslavia led to the establishment of eight more states within Europe, complex diplomacy, military intervention, and war-crimes trials. Genocide in Rwanda and Sudan and state failure in Chad and Somalia, plus major unrest and a new wave of civil wars elsewhere in Africa, have brought much new diplomatic attention and aid efforts to that continent since the 1990s. Beginning at about the same time, the threat of international terrorism rose on the global scene, culminating in the September 11 attacks on US home soil and other terrorist activity on Guam and in Madrid, London, and Mumbai. In response, the United States fought wars against Saddam Hussein's regime in Iraq and against the Taliban in Afghanistan and northwestern Pakistan. The surge in international aid was subsequently augmented by a great expansion in military operations.

Calls for the international community to deal with conflict extended in other directions. International piracy in the Straits of Malacca and the Gulf of Aden required intervention by and coordination among a number of nations. Nuclear politics emerged anew, and included intense negotiations over North Korean nuclear weapons proliferation conducted during the Six Party Talks (the United States, Russia, China, Japan, and the two Koreas); the nuclear stalemate between Iran and the West over Iran's uranium enrichment plants; and the nuclear black market set up by A. Q. Khan, the Pakistani "father of the Islamic bomb." Elevated tensions between Pakistan and India, leading to a renewed nuclear arms race in South Asia, mark another source of hostility with global consequences. The collapse of peace negotiations between Israel and the Palestinians and Israel's brief 2006 war against Hezbollah in southern Lebanon have brought much new diplomatic and media focus to that region.

Global emergency aid agencies have been challenged and impelled to expand their capabilities and reach because of several natural disasters of terrible proportions and impact: the 2003 heat wave in Europe (35,000 dead), the 2004 tsunami in the Indian Ocean (230,000+ dead), Typhoon Nargis in 2008 that struck Myanmar (145,000+ dead), and the earthquakes of 2010 in Haiti (150,000+ dead or missing) and 2011 in Japan (20,000 dead or missing).

A number of other global issues have come of age in this period of time. The United Nations' Millennium Development Goals—to end poverty and hunger, establish universal education and gender equality, improve health for mothers and children, eliminate HIV/AIDS, and promote environmental sustainability—mark that organization's ambition to improve the quality of life worldwide. Likewise, the UN Conference on Environment and Development of 1992 (also known as the Rio Summit) set in motion Agenda 21, a host of new efforts aimed at education, negotiations, and policies related to sustainability and human rights. The conference took up such global threats as deforestation, desertification, ocean pollution, decreasing biodiversity, and not least, climate change. The last of these issues, originating from scientific work performed mainly in the United States and the United Kingdom, has itself become a major theme in global discussions of the future, drawing together other major topics of concern such as water resources, energy supply, and food availability.

Finally, at the highest level, Asia has returned to the center of world affairs in possibly the greatest power shift of the twenty-first century. Focus has most often been on China—not without reason, for its massive program of modernization and its expanded global influence. China's attempt to build an industrial- and a knowledge-based economy in little more than a single generation when it took the West two centuries to achieve both is wholly unprecedented, and has given the country a dominant political presence in the Pacific region and beyond. But the Chinese achievement, and the concerns it has raised, can overshadow other important aspects of the Asian ascent. South Korea, too, has been a spectacular economic and technological success story, particularly since the early 1980s, propelling its own people out of poverty and joining the OECD in 1996. Japan, despite an economic funk lasting over two decades, remains among the world's top five nations by gross domestic product and one of the world's leading scientific and engineering nations. Russia and India are also part of the Asian power shift, rounding out the list of major nations. In the meantime, the next significant wave of modernizing advance has already begun in nations such as Vietnam, Indonesia, and Malaysia.

How have all these events and trends worked to the advantage of spreading the English language? They have done so by eliminating the one politically grounded competitor (Russian) and by advancing the need for global diplomacy, international aid, and arenas of negotiation in which English was already gaining precedence. Rapid decline of the Russian language as a lingua franca in the Baltics, Eastern Europe, and Caspian-Central Asia created a kind of linguistic vacuum that the new nations sought to fill, both with their own indigenous tongues, some suppressed for many decades, and with languages that would provide broader external connections, not only to global markets but to political bodies, technology transfer, and intellectual institutions as well. English, and a few other tongues (Turkish in the Caucasus, for example) were the clear winners here. The rise of Asia on the geopolitical landscape has spread the English language for some of these same reasons, but also because this tongue has acted as a lingua franca within large portions of this region: between India and Pakistan; often among Japan, Korea, Russia, and China, particularly when the United States is involved; and in Southeast Asia (for instance, English is the official language of ASEAN, Association of Southeast Asian Nations, representing ten countries and over one thousand native languages).

Yet geopolitical realities have helped accelerate the growth of English for a series of other reasons:

- the expansion of international peacekeeping, human rights, and law;
- a military coalition led by the United States in the Gulf War, Afghanistan, and Iraq;
- the formation of a protective alliance, via expansion of NATO (a dozen new members since 1991);
- global anti-terrorism measures, particularly following 9/11;
- global finance and economic analysis;
- the worldwide energy industry, including oil/gas sources and organizations such as the International Energy Agency (Paris) and also nuclear power and unconventional sources;
- global rescue organizations, as most of these conduct their operations using English;
- nongovernmental organizations (NGOs) involved in assistance to developing nations, environmental causes, global health, and microfinance; and

- international scientific, engineering, and medical endeavors.

Thus, even before we invoke the phenomenon of globalization and all it involves, we find that English has been promoted by the realities of recent history. In many of these realities, the United States serves as protagonist—but not all, not by any means. America does not dominate aid and peace-keeping in Africa, nor does it lead international rescue efforts. Neither is it the linchpin in negotiations regarding EU policy toward Southeast Asia (which use English); operations of the International Atomic Energy Agency; the International Criminal Court and war-crimes tribunal in The Hague, which the United States has not joined and even opposes occasionally; or the forging of global treaties, such as the Law of the Sea, which the United States had not ratified as of 2012.

Beyond the big players, however, is the panorama of rapid economic and scientific advancement in developing nations. Apart from China and India, these nations are where the most aggressive moves toward modernization are happening. By the late 2000s, they were growing economically at rates more than twice those of any advanced country and were responsible for at least a third of the global gross domestic product. Many of them, emerging from decades of postcolonial struggle, war, and poverty, are seeking to build very different futures. English has a major, if complex, role to play in such efforts, as the situation in Rwanda makes clear.

The Case of Rwanda: English for History, Politics, and Science

In 2008, the government of Rwanda declared English the sole national language, replacing the colonial tongue, French.[17] A year later, the country joined the British Commonwealth, despite its lacking any historical tie to Great Britain. Since the end of the horrific genocide in 1994, English had actually been an official language there alongside French and the indigenous Kinyarwanda. It was declared so after the Rwandan Patriotic Front overthrew the Hutu regime responsible for the mass murder and established a new government. The history here is essential. It shows not only that language is politics, but that English, as a global tongue in economic and scientific domains, has spread by many different types of decisions.

In 1959, the Hutu majority dethroned the ruling Tutsi king, took control of the country, and proceeded to kill thousands of Tutsis and drive many more into exile in neighboring anglophone Uganda and Tanzania. In the years that followed, the Hutu regime built relations with both France and Belgium (the original colonial power). Children of the Tutsi refugees, meanwhile, who grew up learning to speak their native Kinyarwanda and English, formed the RPF, which in 1990 began a three-year war against the Hutu government. Hutu rulers, with support from both France and francophone countries in Africa, fought the RPF until a cease-fire in 1993. Ethnic tension, however, was driven to a near-hysterical level by an extremist Hutu ideology (supported by colonial conceptions of the Tutsi as an ancient invading race) that claimed the Tutsis would kill and enslave all Hutus. Assassination of the Hutu president in April of 1994 unleashed a media storm against all things Tutsi and the mass killings of up to a million people. The genocide lasted several months until the RPF finally gained control. For the new government, English was the language of resistance and victory. French would be necessary for a while; national institutions had operated in this tongue for half a century. But France and the RPF began trading accusations about involvement in the genocide, with Rwanda severing all diplomatic ties in 2006, when a French judge ordered the arrest of President Paul Kagame. Matters were not calmed by revelations that French support of the Hutus was in part a matter of policy, to protect France's waning influence in Africa against "anglophone aggression."[18]

Genocide nearly ruined the Rwandan economy, based in subsistence farming. Yet in the decade between 1996 and 2006, the country experienced an extraordinary recovery; its gross domestic product increased an average of 10% a year. This was possible due to increased aid from Western institutions (the World Bank, International Monetary Fund, USAID [U.S. Aid for International Development]), and government policy also played a role. Under Kagame, Rwanda has sought to improve conditions by raising education levels, building infrastructure, and attracting investment from other African nations. It has done well by expanding trade with English-speaking Uganda, Kenya, and South Africa. It wishes to extend these efforts further through a new Ministry for Science, Technology, and Scientific Research, whose purpose is to put applied science at the center of plans for national development. A research establishment will be erected to help improve crop yields, provide better irrigation and electricity, and offer vaccines and general health care. In 2010, per-capita income for Rwanda was still less than $1 per day; it remains one of the poorest nations on earth. Knowledge-wise, its stated goals could be easily met: the technical know-how not only exists but is at work in other countries. Yet Rwandans are plagued by a lack of electricity—this alone limits any and all efforts at modernization and has also led to rapid deforestation (wood for fuel). In 2010, the power supply never exceeded fifty-five thousand kilowatts (55 MW)—barely enough for a few thousand homes in the United States—meant to serve a population of over 10 million.

An innovative approach to improving power generation began in 2009, led by companies from the United States, South Africa, Denmark, and Rwanda itself with assistance from the government of Congo, backed by expert advice from American energy firms and an aquatic research institute based in Switzerland (Eawag). The approach involves extracting methane gas from the deep waters of Lake Kivu; as much as 2 trillion cubic feet of it has accumulated in the lake from respiration by anaerobic bacteria of carbon dioxide, which leaks into the dense, lower waters from volcanic sources associated with the East African Rift System.[19] The lake's shores are heavily populated, however, and there are possible dangers with the project. Any sudden escape to the surface of both methane and carbon dioxide could suffocate all nearby life, as happened in 1986 at Lake Nyos, Cameroon, where seventeen hundred people died. Lake Kivu, however, is more than a thousand times the size of Nyos, and the number of inhabitants who live close by equally larger. A gas eruption could therefore be disastrous. Wishing to know more about these dangers and to better evaluate the resource, the Rwandan government cosponsored a 2010 symposium, in English, on the detailed geology, history, and hazards of Lake Kivu.[20] Participants came from the United States, Italy, Switzerland, Belgium, Congo, Tanzania, Ethiopia, and Canada.

Rwanda's development, its future economic growth and power supply, and the safety of Lake Kivu's population all depend in some measure on the English language. English is not the guarantee for any of these to happen, of course. Language cannot trump political or social reality, and is more often its tool. But this is precisely the point. Implementing English would be an important element in a larger development plan for Rwanda, necessary but far from sufficient. With only about 10% of the population proficient in this tongue now, the shift to English will likely bring struggle, even resistance, and could easily take generations to achieve. Kinyarwanda will remain the primary national language, not least in education, while English will be taught as a foreign tongue. The quality of instruction and, most of all, the equitable distribution of education itself will be crucial.

Powers of Comparison

Rwanda thus provides a case of how English fits into complex (and, at times, violent) histories and how its global status, not least in science and technology, has fueled its continued expansion, occasionally at the expense of other international tongues (in this case French). Here is where we might return to a topic mentioned earlier: how does English compare with humanity's other major languages?

The estimates in table 2.1, assembled from a number of different sources, are necessarily rough but nonetheless give some help and perspective. They show a possible range of first-language speakers for the world's ten largest tongues, global distribution in terms of the number of countries where they are spoken, and whether each appears to be expanding, relatively stable, or decreasing in degree of international use.

Several points leap from these numbers. Adding up the maximum number of native speakers in this list, the sum (3.2 billion) accounts for nearly half the human population (7 billion by late 2012). The number of nonnative speakers of these tongues is also large—adding native and nonnative yields 4.9 billion to 5.5 billion people, reflecting the multilingual nature of the world. In terms of the most widely disseminated languages, after English comes Arabic, then Spanish, then German, and Portuguese. Obviously, this reflects a very long history of colonialism and empire (Arabic was a colonizing language across North Africa, the Middle East, and Central Asia in the seventh and eighth centuries CE). Indeed, a map of the world's major international languages today will be largely a map of past colonial and imperial possessions (the exception is German, to some degree, whose distribution is also related to emigration in the early twentieth century). Mandarin, meanwhile, is less widespread but actively expanding, due to new immigrant communities in regions like Central Asia and Africa, where Chinese "economic colonies" have been established.

TABLE 2.1 Estimated ranges for native speakers of world's ten most widely spoken languages

Language	*Number of native speakers*	*Global distribution: number of countries**	*Nonnative speakers****
Mandarin	850 - 930 M**	20 - 30 exp	15 - 20 M ʃ
Spanish	360 - 400 M	45 - 50 exp	50 - 70 M
English	360 - 380 M	115 - 140 exp	1,500 - 1,600 M
Hindi	360 - 380 M	20 - 25 sta	120 - 150 M
Arabic	220 - 275 M	57 - 60 exp	100 - 150 M
Portuguese	180 - 210 M	37 - 40 decr	15 - 20 M
Bengali	180 - 205 M	10 - 12 sta	30 - 50 M
Russian	150 - 200 M	33 - 35 decr	75 - 100 M
Japanese	125 - 126 M	15 - 20 decr	<10 M
German	90 - 120 M	40 - 43 decr	10 - 20 M
TOTAL	2,875 - 3,226 M		1,916 - 2,190 M

Sources: *Ethnologue: Languages of the World* (Dallas:SIL International, 2009)

David Crystal, *English as a Global Language* (Cambridge: Cambridge University Press, 2005)

David Graddol, *English Next* (London: British Council, 2006)

R. E. Asher and Christopher Mosely, *Atlas of the World's Languages*, 2nd ed. (London: Routledge, 2007)

Keith Brown, ed., *Encyclopedia of Language and Linguistics*, 2nd ed. (New York: Elsevier, 2005)

Andrew Dalby, *Dictionary of Languages* (New York: Columbia University Press, 2004)

Bernard Comrie, *The World's Major Languages*, 2nd ed. (London: Routledge, 2009)

Jeffrey Gil, "A Comparison of the Global Status of English and Chinese: Towards a New Global Language?" *English Today* 27, no. 1 (March 2011), 52 - 59.

*Using the language as main or important medium of communication, oral or written, whether in society at large or sizable immigrant communities.

**Includes spoken and/or written.

***Includes estimated number who employ each language for actual communication. This leaves out students studying the language in school and those only able to recite lines from songs, films, religious texts, and other such non-interactive uses.

ʃDominantly Chinese with a non-Mandarin tongue as their first language, e.g. Hakka, Wu, Yue, Uyghur.

Abbreviations for overall degree of international influence:

exp = expanding

sta = stable

decr = decreasing

Table 2.1 lacks one important language, with forty-five to fifty nations in its community—French has been declining, not only in Africa, but also in Southeast Asia, where it is being replaced by English and Chinese.

English stands in a different category from all other tongues. It shows that the number of native speakers is not so important in terms of global spread and influence. The 900-million-or-so speakers of Mandarin are overwhelmingly Chinese; so are its nonnative speakers, whose mother tongue is one of the other important languages in China: Wu, Yue (Cantonese), and Minbei. This is also the case for Hindi and Bengali in India and Bangladesh. Russian is an interesting and complicated case because of the collapse of the Soviet Union; many millions in former Soviet republics are either ethnic Russians or know Russian partly as a native tongue, partly as the tongue of a conqueror/occupier, and partly as the tongue of the largest nearby economic and political power. After English, the most studied language worldwide is Spanish, which may well have on the order of 50 million to 70 million nonnative learners.[21] Far smaller numbers are given for Japanese and German, despite their wide extent in the first half of the twentieth century (German continues to be spoken locally as an additional tongue in parts of Eastern Europe and in Luxembourg, Lichtenstein, Switzerland, and North and South America). Again, French is strikingly absent from our list; once the ruling lingua franca of eighteenth-century Europe and imposed over large portions of West Africa and Southeast Asia, it has progressively dwindled in total extent. English, however, stands beyond all others by an order of magnitude. As our British visitor to the Digum school realizes, native speakers have become a small, shrinking minority.

Geographic distribution leads us to ask about languages used on the Internet, a vast new linguistic space that is itself expanding every year.[22] The question is crucial, since this space is now the medium in which the greatest volume of language exchange takes place worldwide. Based on the number of users, the 1990s saw the United States and the English language dominate the Net. At that point in the Internet's existence, technological barriers didn't allow the easy display of non-Romanized symbols, thus preventing the use of languages from the most populous places on Earth: China, India, Arabic-speaking nations, and so forth. Over 80% of Internet use was in English, leading many to proclaim (and lament) that a linguistic hegemony had already hardened. A decade later, we know this was temporary. The birth of a linguistic hegemon on the Internet had been greatly exaggerated. An updated Unicode Standard character-coding system and the spread of cheap computers and online access during the 2000s altered the landscape. By 2010, Internet use had grown enormously, and English as a percentage of total users had fallen. In 2011, Internet use in English stood at just 27%—still the largest in the world but now less than the combined total of just two other languages, Chinese and Spanish. In fact, during the decade 2000–2010, the rise in Internet usage on the web showed Arabic growing by 2,500%; Russian, 1,825%; Chinese, 1,277%; Portuguese, 989%; and Spanish, 743%, with English at 281%.[23] The meaning of these numbers seems clear: the rest of the world has been catching up. Out of a total 1.97 billion Internet users estimated for 2011, over half were in developing nations; yet this represented a mere 17.5% of the total population in these nations, suggesting that the "digital divide" remained a distinct reality, but also that tremendous future growth in Internet use is likely.

What about content? A common rumor still has it that 80% of information on the Internet remains in English. Yet this percentage has no real verification and counts as hearsay. In 2003, *The Internet Encyclopedia* provided a figure of 68.4%,[24] a rather precise percentage. If we focus on the web alone, minus social media, could this be close to accurate? One hint: algorithms designed to identify and rank web pages by frequency of use (such as HITS, PageRank) still overwhelmingly identify English-language sites. Another indirect indication: the number of actual direct English users in 2011, according to the website Internet World Stats, was 565 million, well over the total global figure for native speakers. Thus, large numbers of nonnative speakers are searching for and, in many cases, adding to Internet material in English. Such is not the case for Chinese web users (509 million in 2011), nor for those who use Spanish (165 million) or Arabic (65 million).[25] Middle Easterners are not adding new content in Mandarin; those in China and Spain are not putting up sites in Arabic. They *are* doing these things in English, however.

Much will change, and not change. Digital technology, from netbooks to cell phones, is far from universal; the World Wide Web is not at all worldwide. There is a great distance to go before people in every country are wired, computer owning, and computer literate. In 2011, no fewer than 1.5 people were living without electricity, and it is forecast that over a billion will still be without it in 2030.[26] Nearly half the world's population continues to live on less than $2.50 a day.[27] The digital era has not even begun for most of humanity. Internet use may be soaring—between 2000 and 2011, the number of people online increased from 500 million to 2 billion—but economic and energy realities will have much to say about how that growth continues and which new linguistic communities find life online. Social media promise much new opportunity. Such is how Internet use has expanded most rapidly in recent years, particularly in poorer nations.[28] There is the possibility that many native tongues could be brought online. Yet there is just as much of a chance that this won't happen, that the needed organization, financing, and will won't come together, because speakers see better prospects in a more dominant language.

For now at least, English is the only language on the web whose regular users extend far beyond the bounds of its home countries. According to Kiyo Akasaka, the UN's under-secretary-general for communications and public information, English-language websites in 2009 continued to dominate the Internet at about a 70% level.[29] Again, this statistic may exaggerate the case—it is well known, for example, that businesses of all types and sizes practice "web localization," ensuring their sites are in the local language(s) wherever they are trying to sell goods and services. Yet if these businesses have outlets in touristed areas, in local financial centers, shipping ports, travel hubs, and university towns, they will have an English version of their website, too.

Among my own students from East Asia, nearly all use websites in English outside class, even though their skill in the language varies. Most view English as a required tool. Popular culture and social networking sites are major draws, but a few students have created web pages in English as part of their work for a government agency or an international company. Chung-Hee Park from Seoul told me,

> At home, in university, we use English websites because we must. I am interested in steel industry. Some of best companies are in Japan and now China. But Japanese companies, like Japan Steel Works, who makes world's best nuclear reactor vessel, have all their websites in English. Chinese companies are doing this too, more slowly. I can find other English sites with information about Angang Steel, for example. It's much easier than learning Chinese.

I ask about website translation, like that offered by Google and Babelfish. "Terrible result!" he says. "Why use it? The website is already in English."

State of the World's Languages: Under Threat

Any comment comparing the extent of major languages today will raise the matter of the general state of humanity's linguistic endowment. The blunt situation is this: 6,909 tongues have been counted (by *Ethnologue*), and no less than 94%–96% of them are indigenous languages. They are spoken by communities with between 1 and 10,000 members, who in total make up less than 6% of the global population. In brief, the diversity of human language rests overwhelmingly with small groups of native peoples scattered among nations dominated by much more widely spoken and socioeconomically powerful tongues. In a modern, textual world, these 6,000-plus native languages are therefore desperately vulnerable.[30]

Still more, a great number are in peril. Half are already described as "moribund," meaning that children longer learn them. The rate of their extinction has been accelerating and may well eliminate the greater portion of them before 2100. About this there is little debate. Some of the figures, including those given above, are dramatic. Another striking fact is that only 200 to 300 languages are actually written—meaning that they have a full-scale orthography, or writing system, in daily use by the relevant speech community. This figure is an approximation at best, but appears to be correct.[‡] Needless to say, orthographic languages are the only ones that can be used for literacy, education, the professions, and general communication with the outside world, as well as access to the Internet. North America in 2010 had about 194 living tongues, compared with well over 300 before colonial settlement. But of the total today, a mere 33 are spoken

[‡] Portions of the Bible (e.g. the Lord's Prayer, Luke 11:2–4) have been translated into more than 2,000 tongues, including many Native American languages such as Arapaho, Comanche, and Hopi. All but a few hundred of these are indigenous and are otherwise entirely or dominantly oral. Their translations are thus phonetic, using the alphabet of a regional orthography, such as Cyrillic or Roman, to inscribe the spoken sounds for each biblical passage. This is certainly not the same thing as having an entire writing system.

by both adults *and* children.[31] Such provides one idea of how desperate the situation really is in some of the more linguistically rich areas. As the world shrinks in the digital age, its diversity of speech withers.

All this leads to a potent conclusion. People in the first few decades of the twenty-first century may be the very last to live among a major part of humanity's linguistic heritage. It would seem to be a pressing goal, then, to use digital tools to document what still exists as well as what has already been lost. Clearly, a global archive along these lines would be a profound and fertile necessity. Some early steps have been taken in this direction. The United Nations Educational, Scientific, and Cultural Organisation (UNESCO) has assembled *Atlas of the World's Languages in Danger,* which now includes 2,500 entries.[32] *Ethnologue* mentions 473 tongues worldwide at the very edge of extinction, mostly in the Americas—Brazil, Colombia, Peru, Canada, and the United States. Australia has over 90 aboriginal tongues that are now moribund. Papua New Guinea and Eastern Siberia are also "hot spot" areas. Rescue efforts are being made in all these places.[33] There is a growing idea of what could be lost, not only linguistically but culturally, historically, and also scientifically, in terms of native knowledge about local ecologies, medicines, and other wisdom.

We might therefore hope that any documentary and rescue endeavors will find the support they deserve. In some cases, this is happening. North America, for example, has seen dozens of indigenous tribes with income from profitable casinos hiring linguists and teachers to preserve and, through local schools and media channels, revitalize their native tongues. Elderly speakers still fluent in the language, who grew up speaking it, become invaluable "reference books" in many of these efforts. Yet these people can be shy to come forward, partly because of the many decades when their native speech was treated as a mark of backwardness in comparison to English. Preserving these languages is a race against time, in other words, yet cannot escape the scars of history. A number of tribes have made sizable, multiyear grants to university departments of linguistics with the goal of creating dictionaries, grammars, oral histories, and other documentary material that will help keep their languages intact and perhaps even aid in creating writing systems for them.[34] These are all laudable efforts—efforts not only to preserve forms of speech but to reclaim specific identities. That they will succeed isn't certain, but the odds are enormously improved from a few decades ago. That they are taking place in the richest nation on earth, however, may not be an accident.

That so much threat should be happening in modern times makes intuitive sense. Colonialism did much to disperse and destroy indigenous peoples, hence their languages. In Europe, the rise of the nation-state mandated a single unifying language (or a very small number of them), marginalizing local tongues. Today, too, dominant national languages are forcing native ones to be abandoned worldwide. Spanish in Latin America, Russian in Siberia, and Arabic in North Africa, as well as English in the United States, Australia, and Canada, are taking a toll. The tale has a few bright passages, in that successful rejuvenation has occurred—Welsh, Irish Gaelic, Breton, Maori. France's constitution was amended in 2008, granting official recognition for minority tongues. Yet these victories can't hope to compete with the thousands of languages that have already vanished or are threatened today.

Languages become threatened when people stop speaking them. They do not literally die; they are not living creatures, cultural "species" or whatever, but systems of communication that can be abandoned, replaced, or cast into archival preservation. (The Darwinian view of languages as "species" is effective at the level of metaphor, but misleading when taken literally, in terms of "fitness," "selection," and so on, suggesting that linguistic change is part of nature and unrelated to historical events.) A number of reasons may work together toward this end:

- *invasion or war*, when the languages of a conquered people are weakened or replaced by the invader's speech, which becomes the dominant tongue;
- *famine, disease, natural catastrophe*, any of which can enfeeble a language community;
- *environmental change*, such as conversion of tribal land for agriculture, deforestation, or desertification, causing a people to leave;
- *migration*, displacing a linguistic community (or a large part of it) that is then assimilated into a new language setting;
- *urbanization*, with people from native communities moving to cities and adopting the dominant culture and language;

- *culture shift,* whereby members of a native community (especially its youth) choose to abandon traditional ways and adopt modern lifestyles and speech;
- *language policy,* where a government restricts the number of tongues to be used for education or in various domains; and
- *parental decisions* to have children learn nonnative languages for their advancement, thus confining a native tongue to purely vernacular settings.

What does the spread of English have to do with all this? It has been blamed as the primary agent for bringing language endangerment to its current high level. As the tongue of globalization, it has been accused of "linguistic imperialism."[35] Are American and British governments and companies colluding to employ this language as a means of economic colonization? Has it achieved such power that it now acts to enfeeble native vernaculars everywhere? Has it become therefore not only a threat but an actual "killer language," capable even of "linguistic genocide"? Such questions, which draw from a portion of the scholarly literature, show that the line between metaphor and literalism has been crossed, or forgotten. Fear and loathing aimed at the English language, though not entirely rational responses to language dynamics today, do reflect how overwhelming the impact of this one tongue can appear, and how helpless other societies seem to be in the face of its expansion. We are thus compelled to look at the evidence.

Empirical studies offer four main points. First, the global spread of English is a very uneven and variable phenomenon. In some nations, it is widely spoken and used, as in Scandinavia. Yet in other countries such as Japan, where it is taught to the vast majority of children, its use outside the classroom is minimal, and it is the national language that threatens local native tongues (in Japan's case, Ainu and Okinawan languages). In such instances, English competes against other *international* tongues. Any threat to a native language happens when people decide or are forced to adopt English as a *vernacular,* something that has clearly happened in some places, such as North America and Australia, but that is far from a global rule. Second, English is by no means the only international tongue relevant to language death. Regions with higher threat levels include South America, Siberia, eastern India, and Papua New Guinea, places where local vernaculars are giving way to Spanish, Portuguese, Russian, Bengali, and Indonesian.[36] Third, when it comes to their speech, people are not uninformed, routine victims. In some cases, they are forced to give up their native speech. Yet more often, they are active agents making decisions they believe are in their own best interest and in the interests of their children. They may well deserve "linguistic rights"; they are also core actors in the drama of English.[37]

What forces are most active in language threat today? Field studies suggest that urbanization, environmental degradation (affecting small, tribal communities), culture shift, and parental choice rank high. A few of these causes are indeed linked to globalization; some are not.[38] When we put them together, however, several factors come immediately forward.

Children are the most critical linguistic resource, in every case. They are the link to every future, the source of survival for any language. Thus, whatever impacts their language learning is also crucial. *Education*, then, must also be a core factor. Languages taught in school have a better chance of gaining or maintaining broad use. An education system must choose its tongue(s), and this is a political decision. One idea—to train children in both a dominant tongue and their native tongue—seems excellent, but isn't trouble free; choices among local languages must still be made, and teachers trained. Education is no fix for language conflicts, and often is a battleground itself. Putting education and children together suggests how important *perceptions of opportunity* are. Here is meant not merely higher income, but the chance for more rewarding work; better living standards; improved health care and food; a safer environment; access to modern forms of entertainment and recreation; intellectual material; and political participation. All these aspects, finally, are concentrated in *urban areas*, the future of humanity.[§] Cities are the centers of modern life: government, the economy, universities, businesses, the mass media, youth culture, and much more all live here, where the greatest number of languages rubs shoulders in the smallest space. Consider New York City, where no fewer than eight hundred languages are believed to exist,[39] many of them endangered in their native areas, but less imperiled here due to use among small groups.

[§] In 1950, under 30% of the world's people lived in cities, but by 2009, 50% did, and by 2050, it will likely be 70%. This is a profound change from the past, when rural life based in agriculture ruled human existence since Neolithic times.

Meanings

The children at Digum School reveal that for a great many who will never see the United States or Britain, English is nothing less than the tongue of modernism itself. It speaks not only of the possibility for personal and national economic development, plus alliance with (and aid from) the world's superpower; it is also attached to ideas such as progress, internationalism, and emergence from poverty. Many poorer countries and communities regard English as a medium of global connection and advance. "I do not come here to be American," says Daniel, a student (of mine) from Kenya, "but to be a better African, a more sophisticated person of my country." Learning English thus has symbolic as well as practical dimensions. If it is felt to be a needed element in any program to produce trained diplomats, businesspeople, scientists, and physicians, it is also a carrier of what it means to be cosmopolitan, worldly.

In a majority of cases, English is not forced on a people—in the manner, say, that Castilian Spanish was forced on the people of Spain under Franco or Japanese on the population of Korea in the pre–World War II era. It does not bestride the world as a tyrant, always looking for new victims and minions. Rather, it is chosen by governments, parents, professionals, and young people, usually for the reasons noted above. In some cases, this choice acts to impose a language, directly or indirectly. Yet it is also true that governments, companies, schools, and professions must make choices, and even when they do so with linguistic diversity in mind, circumstances turn out to be more complex than anticipated.

In South Africa, for example, linguistic policy grants equal privilege to eleven tongues, including nine major native languages, yet without any specific, legislated guidelines on where and how they should be used, so that the status-quo momentum toward English, with the educational system behind it, continues. Communities in South Africa generally feel that this language can gain more opportunity in more areas, and so parents choose it actively, at a grassroots level. The situation is complex; bi- and multilingualism among students is common. The country's national anthem is a mixture of all eleven official tongues, with different stanzas spoken in each language—a unique and not wholly convincing or successful attempt at symbolic integration. There is lingering resentment among blacks (80% of the total population) toward learning Afrikaans, the tongue used by apartheid governments and still widely spoken among the white population. Afrikaans must be counted as a native African tongue (and Afrikaners have always considered it so). Its speakers today, in their turn, may well resent the spread of English, as a tongue imposed by the British on the "native" Boers ("white-on-white" colonialism). English is indeed rapidly replacing Afrikaans in professional domains, though spoken in only 9% of homes and only 1% of homes of black families.[40] The whole matter defines a long-standing series of political-historical struggles still waged at the level of language, but not controlled by any entity. It is perhaps ironic, though not at all surprising, that South Africa has developed its own variety of English that reflects this complexity, having absorbed vocabulary from several indigenous tongues, including Zulu and Bantu, plus many words from Afrikaans, some of which are now part of Anglo-American English as well (*apartheid, trek, veld, commando*).

Localized, or rather nativized, varieties of English that differ significantly from one another in their lexicons, pronunciation, and even grammar exist throughout the world, as noted in chapter 1. World Englishes, which have adapted the language to new cultural and linguistic places, mainly occur—thus far—in former British colonies and cannot be counted as mere dialects or as wanderings from some (mythical) standard of purity. They represent a dynamic aspect to global English, an aspect that has characterized all tongues which have spread widely across various nations and peoples.

A powerful point: world Englishes are evidence that Anglo-American speech has retreated as a final linguistic model. In terms of discourse generally, English has proved itself extremely flexible in expression and use. To hear a Liberian pastor say to his congregation "Come forward with dancing on your feet" or a Hong Kong bookseller to a customer "I am strongly recommend this book to you," may suggest poetic license or quaint error, but time spent in either place will show that both are accepted forms. Some more examples:

He speaks chaste Hindi. (India; meaning "pure Hindi")
My friend would like to become a navy. (Nigeria; wants to join the navy)

Residents will repair the roofs on a gotong-royong basis. (Malaysia; cooperative basis)
That accident was happened at 6 p.m. (Hong Kong)
Chicken here is always *tok kong*. (Singapore; always delicious)
Pound a fret can't pay ounce a debt. (Jamaica; worrying can't reduce a debt)

Anglo-America, in other words, has supplied starting material that the rest of the world has been busy shaping and making its own. As we will see in the next two chapters, this has its inevitable reflection in formal writing, where new levels of rhetorical flexibility have come into play—a reality that has begun to impact the sciences.

Varieties of English have not developed rapidly or smoothly. The idea that this tongue (or any other) can be implanted successfully in just a generation or two, even in a few select professions—a fairly common hope among a number of nations and their educational systems today—has little historical backing. Reasons have much to do with cultural politics, as well as the unintended consequences of any language policy. Take the case of India. Here the language has been taught since the early nineteenth century, yet has penetrated only 10%–15% of the population. English was initially withheld, for both control purposes and the belief that education in native tongues was best for a people unready to absorb European culture. By the late eighteenth century, Indian leaders argued for the learning of English on the grounds of economic progress and modernization.[41]

The decision to shift policy came as a result of the famous "Macaulay Minute" of 1835, written by Lord Thomas Macaulay, a humanist on the Supreme Council for India. He argued that education should help "to form a class who may be interpreters between us and the millions whom we govern; a class of persons, Indian in blood and colour, but English in taste, in opinions, in morals, and in intellect." Science was to be a core part of this—the new class had the task "to refine the vernacular dialects of the country, to enrich those dialects with terms of science borrowed from the Western nomenclature, and to render them by degrees fit vehicles for conveying knowledge to the great mass of the population."[42]

History caused damage to Macaulay's plan. English education did flourish, but while British policy recognized Bengali, Hindi, Urdu, Guajarati, and other native tongues, schooling in them received little backing and languished as a result. Demand for English grew in urban areas, where job opportunities in the colonial bureaucracy were plentiful. Many students in the new English schools were from poorer families and studied only to gain clerical jobs. Their language skills were limited; their curriculum, taken from England's primary schools, was ill suited to second-language speakers. Better tutoring existed for middle- and upper-class children, destined for managerial work. English thus enforced existing social inequities even as it expanded opportunities for some. Irony, however, interceded in another way. By the early twentieth century, the class of poor English speakers had absorbed ideas of independence and self-determination from the very education intended to keep them colonial subjects. Such became an element in Indian nationalism, also pursued by a segment of elite English speakers (such as Mohandas Gandhi). The English language, therefore, worked both against and for India's own eventual independence. Its distribution was always limited and restricted. Its meaning, including symbolism, was always mixed.[43]

India thus provides not a model but an indication. Importing a dominant tongue to a society usually means adding a new medium of power to a setting populated by various forms of inequality. Such a tongue can thus be used to enhance inequity or reduce it. Guiding the use of this language for the latter purpose has never been an easy task.

The Question of Standards

The reality of world Englishes raises the issue of standards. Are there now no final reference points for the English language? Are such standards soon to become irrelevant? The answer is no, as any corporate manager or scholar will tell you. Anglo-American English has not lost its role as a global norm in one major and critical domain: professional *written* discourse. Here, what is called Standard Written English (SWE) continues as a de-facto world orthodoxy. Is it, however, an absolute norm?

Again, the answer is negative. SWE, after all, is a moving, evolving norm. Defined as that version of the language most widely accepted as "correct" or "proper" or "acceptable," it is determined not by a unique authority or group of powerful nations but instead by a consensus of gatekeepers—publishers, editors, educators, authors of language textbooks, speechwriters, journalists, and other middlemen who actually shape the form of the language that reaches a reading public. Given these varied agents, working in different types of professions, different nations, and also at the international level, it isn't a shock that SWE never achieves true uniformity. Yet aside from some truly minor differences, the norms of professional writing in the United States and United Kingdom, Canada and Australia, New Zealand and Ireland do approach a working standard, one that is greatly reinforced by a constant stream of publications from international bodies including the United Nations (on population, treaties, climate change, urbanization, and so on), EU, World Bank, World Health Organization, and many NGOs.

Yet SWE does not rule everywhere. World Englishes have an influence in this domain. Consider, as an example of East African written English, the following passage from a formal speech by the vice chancellor of Makerere University, Uganda, welcoming international visitors to a conference in 2011 on engineering advances:

> This Conference has been organized by Makerere University, College of Engineering, Design, Art and Technology under the theme Contribution of Scientific Research in Development. There couldn't be a better time than the present to bring together scientists, researchers, professionals and industry leaders from all over the world, to share experiences in advances in technology to transform lives of the population of the world with ever increasing challenges. Makerere University is proud of her academic units like the Faculty of Technology . . . that has kept research and dissemination of research high on the agenda. As you might all know, carrying out research and disseminating it through such Fora is one of the major roles of Universities in the quest for creating new knowledge. The Faculty of Technology in addition has been carrying out applied research in various areas including renewable energy, vehicle design, data management systems, e-labs, low cost housing, and many others. Participants are encouraged to pay a visit to these innovations and research centres in the university for appreciation and knowledge sharing.[44]

A second example, from a report on the state of agriculture in India helps to emphasize the point via South Asian English.

> Agriculture sector is vital for the food and nutritional security of the nation. The sector remains the principal source of livelihood for more than 58% of the population though its contribution to the national GDP has declined to 14.2% due to high growth experienced in industries and services sectors. Compared to other countries, India faces a greater challenge, since with only 2.3% share in world's total land area, it has to ensure food security of its population which is about 17.5% of world population. This leads to excessive pressure on land and fragmentation of land holdings. Against the backdrop of the burgeoning population's demands for foodgrains, degrading natural resource base, emerging concerns of climate change and other challenges, the Department of Agriculture and Cooperation (DAC) has focused on mobilizing higher investment.[45]

Departures from SWE are evident in these two passages—an eighth-grade English teacher in New York City or an A-levels instructor in London would find many details to correct: missing articles, use (and nonuse) of commas, choice of prepositions, syntax. Yet both writings are by adult professionals and have been through a review, editing, and revision process. They each meet existing standards for East African and South Asian English, respectively. Moreover, though not SWE, they are fully understandable to any SWE speaker or writer. They do not incorporate terms from local native languages and do not employ sentence structures or expressions that would be impenetrable to SWE users. The formal, professional context here works to restrain the amount of nativization each non-SWE variety of English accepts. The passages are easily understood by, though clearly not a form of English belonging to, Anglo-America.

The spoils of the battle for "whose English?" will therefore always be intelligibility. To this point, world Englishes have remained fully intelligible to one another in professional writing—a good thing for any profession having a global reach. Yet rhetorical flexibility, as just shown, reveals that each variety of English continues to evolve in this sphere also. There are strong implications, not to mention questions, even concerns, that will need to be dealt with at some point as this evolution continues. Could professional English open itself further to other modes of expression? Could it even accept other styles of scholarly argument, more indirect ones, for example, typical of some Asian forms of discourse? Could professional

English, in other words, become still more plural in rhetorical as well as grammatical flexibility? To some, no doubt, this would qualify as an open-door policy to linguistic anarchy. To others, it might be deemed a matter of linguistic justice.

In practical terms, gatekeepers of written English, to keep global intelligibility intact and primary, will have to restrain future change beyond a certain limit, defined by experience (that is, transgression). This is simpler than it sounds. No organization or authority has full command over English in any of the major professions. No official body exists to create consensus and generate standards, and history suggests such an entity would fail in any case. Even within SWE, there are advisory groups only, such as the Council of Science Editors, and many writing guides for "proper usage." So consensus on standards, however loosely defined, will likely remain in the hands and pens of many and will therefore be uncertain.

The issue of rhetorical flexibility may prove more interesting—and challenging. There is the degree to which different discourses create different varieties of the same knowledge—plural knowledges, in other words, with each variety an identity marker for its immediate community of producers and users. Beyond a certain point, the appeal of such local knowledges would obviously work against or compete with the goal of global intelligibility. A balance, born of tension and experimentation, would then have to be found.

Alternately, there could emerge in the future, at least for some professional fields, a global variety of professional English that incorporates or allows for elements from many local discourses—a kind of broad, forgiving world standard. This may appear to be less likely for now, as world Englishes continue to evolve on their own. Yet in some areas, such as the natural sciences, where a high degree of global consensus comes to exist for established knowledge, a set of flexible norms roughly agreed upon by journal editors, publishers, and researchers appears to be possible. Growing fragmentation (and thus isolation) would be resisted; it would be seen to generate "rogue" knowledges and thus not serve the international scientific community well.

Another interesting matter is how far world Englishes might expand. There's no real way to tell, for instance, how many new varieties might develop in future decades. Nor can we determine, at this point, how the spoken form of each variety might continue to individuate itself away from the others, thus how fractured English's global lingua-franca capability might eventually become at the level of speech. Some international tongues of the past (Latin, Arabic), nativized to different cultural-linguistic settings over a wide region, eventually broke apart into separate, mutually unintelligible tongues or varieties. Others (Greek, Persian), though they continued to evolve, did not. Thus, we have no clear guide. Nor can we say that if varieties of English continued to develop and to grow apart, this would *necessarily* liquidate global intelligibility for professional written English. We can gain important hints and clues from a look at international languages past, as we will do in a later chapter. But there are no real precedents, no replicas, no tablets with engraved rules to tell us what will be. There has never been a global tongue before.

Circles of Speakers—and Learners

Keeping such examples as Rwanda and India in mind, we might ask whether there isn't some way to model the growth of English in a global context. In fact, linguists have attempted to do so in visual form several times over. Given the geographic dimension involved, this can be quite helpful. Most models employ the standard, basic categories of English as a native language (ENL), second language (ESL), and foreign language (EFL). We know these are rough and rude approximations, their borders overlapping and porous. So another way has been found to represent the world(s) of English.

Originally proposed in the 1980s and later updated by its originator, Braj Kachru, it depicts global English as three overlapping circles (fig. 2.1).[46] An *inner circle* includes the main the anglophone countries, where the language remains the dominant mother tongue. A larger *outer circle* consists of countries where English approximates a second language, implanted mainly through British colonialism. Finally, a still larger *expanding circle* involves all those nations where English has more recently been introduced as a foreign tongue (that is, nations having little or no previous historical connection to it), as in Rwanda. Kachru also added another dimension. He spoke of the inner circle as being "norm providing"; outer-circle countries as "norm developing" by modifying inner-circle standards to form new

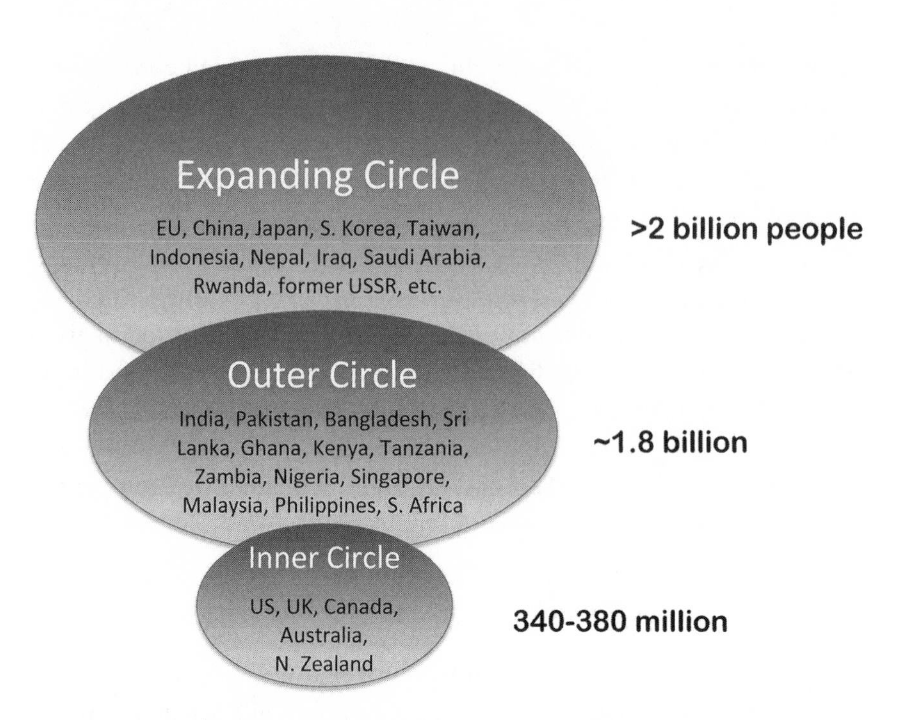

FIGURE 2.1 Three-circle model of countries where English is spoken and being learned. Numbers indicate total population in each circle, not total English users. Adapted from Braj Kachru, *The Other Tongue* (1992).

Englishes; and expanding-circle nations as "norm dependent," relying on the other two circles for standards, hence not (yet) creating their own nativized varieties of the language, but soon able to do this.

Thus, figure 2.1 aims to model several things at once: the overall scale of English use worldwide; the historical-geographic pattern of the English diasporas; the general size distribution of total populations in English-using groups (expressed in comparative circle size); and even the mixture of relative independence and overlapping dependence of each user group. This is a lot to ask of any diagram. Not surprisingly, it has invited a swarm of criticism as well as endorsement. In fact, figure 2.1 has become something of a standard itself in discussions of global English. Be that as it may, probably the most common complaint about it involves the historical inaccuracy of the boundaries between each circle: a large and growing number of outer-circle users now have English as one of their mother tongues, and people in expanding-circle nations have become ESL speakers in significant and increasing numbers. Other criticisms have noted that the phenomenon of world Englishes weakens the interpretation of inner-circle nations as "norm providing," since outer-circle speakers are now doing this, too, ever more often—teachers of English in East and Southeast Asia now include people from India, Sri Lanka, and Singapore, for example. Other critiques do not like that the diagram suggests uniformity among all nations within a particular circle; placing China and Rwanda in the same category makes some scholars uncomfortable, given the differences in development level, economic capacity, and interest for English among the respective populations. What all these criticisms share, then, is a request for a more refined portrait, to reveal the true complexities involved.[47]

If we are less demanding of perfection, and ready to forgive some of Kachru's more ambitious aims, figure 2.1 does offer a helpful beginning picture and also contains an implication that may have been overlooked. It implies that one mark of a global language is to be denationalized, but then also to be renationalized or regionalized, in country after country, as EFL blurs into ESL and possible ENLs. This dual movement really offers a restatement of what we have been discussing—that English, as a global communicational currency, belongs to everyone who uses it, not least those who eventually adapt it to their own setting and needs.

Every part of figure 2.1 depends on a fundamental factor: teaching. Wherever learning English occurs, the teacher steps into the spotlight. Talk as we may about global political and economic developments, the growth of cities, and the ambition of parents, a big part of the future of English and its relation to other languages comes back to the basic interaction between students and instructors—the who, what, where, and how of classrooms. This is just as true of the inner circle as the outer and expanding circles, of course. But it is the latter two that create the possibility and reality of a global language.

As it happens, language learning is also a domain of much new debate and rethinking. Scholars, educators, and officials have all begun to agitate for newer models in English teaching. Now the widespread claim is that the traditional native-speaker model is too narrow and unhelpful. It does not really profit a Ugandan or a Vietnamese student to learn English through the norms of American or British culture, being fed images, vocabulary, and idioms, for example, attached to suburbia, supermarkets, and Samuel Johnson. Such teaching makes sense if we are linguistic imperialists with designs to promote US-UK culture as a golden standard. Questioning the mother-tongue speaker as required standard for fluency will appear to be absurd to some. What other standard *is* there? How can real competence be measured if not by the example of a first-language speaker? Yet as a global tongue, English is no longer umbilical to the United States and the United Kingdom, acting far more often as a medium of communication between people from non-anglophone nations and cultures, for both international and local needs. Actual experience with today's English learners across the world again strongly suggests that inner-circle countries are not the universal provider of norms they once were. Business, scientific, and diplomatic professionals who travel often testify that an enormous amount of daily communication in English takes place, with skill and efficiency, among those who do not have, or even approach, Anglo-American-type speech. Moreover, since varieties of the language now exist, it makes little sense to employ a single standard, native to only a very few nations, in all teaching. Does it really make sense to try and train a child in Ethiopia or Panama to speak like a native of Oxford or Iowa? The native-speaker model (read: US-UK standard) prevents people in non-anglophone nations from taking ownership of the language and the methods of teaching it. Indeed, this model does have its origins partly in colonial ideology—recall Macaulay and his proposal to form "a class of persons, Indian in blood and colour, but English in taste, in opinions, in morals, and in intellect."

As a global language, English can be taught, and more effectively learned, by using references to a student's own daily experience and culture. Such is the new view.[48] Clearly, there is much logic in this, given the role and use of English worldwide. Research that follows up on classroom results confirms that very few nonnative speakers ever come close to the competence of a well-educated native speaker—nor do they need to. A high level of skill that allows a person to function perfectly well in the domains where s/he is active, whether in research or daily trade, has proved to satisfy all needs in a great majority of cases. High-level, functional capability, rather than flawless native competence, becomes a far more realistic, achievable standard.** It also helps support the status of local teachers, who no longer have to try and sound like a Brit, Yank, or Aussie. Note these remarks by a well-respected author and teacher in the ELT field:

> Adopting a native speaker model and then hiring native speakers to model it simply serves to let the students know that the model can only be attained by people who look and sound very different from themselves. This also carries the clear message that teachers who do look and sound like them are unable to produce the required model. . . . It's hard to conceive of any other field in which the learners are implicitly informed that, not only can they never achieve the goal that the curriculum has set for them, but that this goal is even beyond their own teachers.[49]

For now, students still want to sound like Americans (most of all), and administrators still often hold faith with hiring first-language speakers. Yet there are obvious limits here. To state one: growing demand for English teachers worldwide, prefaced on 2 billion learners, requires that we empty the anglophone world of its college graduates by 2020 or before.

Understandably, then, local teachers are the future of English teaching. In a growing number of countries, this is already true. Large parts of Europe, including Scandinavia, Germany, and the Netherlands, where strong English speakers abound, employ their own instructors. If English is taught in an ever-growing number of school systems around the globe, a greater number of people will become competent in the language, whether they ever set foot in England or America or not. It may be only a matter of time, therefore, until the prestige granted inner-circle speakers begins to erode. History, after all, is not without a sense of irony. The real casualty from the global spread of English may well be the native speaker himself.

** This isn't a wholly new idea, by any means. For several decades, there have been courses that teach such things as "business Japanese" or "scientific German." But extending the idea of functional competence to an entire foreign-language program (with options, perhaps, for higher levels of skill, too) marks a change from the past.

The rest of the world will be bilingual at minimum, while the once prestigious Anglo-American speaker will be marooned in monolingualism.

Summing Up

English use worldwide far exceeds anything in recorded history, and continues to evolve in multiform and complex ways. This reality has brought profound reevaluations by the linguistics profession about the nature and evolution of language as an international phenomenon, how its speakers and learners and teachers need to be understood. The diverse functions that English now commands, spanning popular music and high-energy physics, and its dominance at the international level in so many fields of professional and scholarly pursuit, almost guarantee continued expansion. Support for this conclusion also comes from data on international student mobility, a key indicator of national and language status. Yet there are no real precedents to the power and place of English today. Any predictions about the future of this language must be offered in tentative fashion.

Historical factors have been key to the spread of the language in all its varieties, and they remain so. Such factors have recently included major geopolitical events, such as the end of Soviet communism; economic expansion and globalization; concerns over energy supply; international efforts to address world problems such as poverty, disease, loss of biodiversity, climate change; and also natural and manmade disasters. All these developments have a direct connection to the sciences as well—indeed, they do so largely *because* of the global role that English now plays in every scientific discipline. To take but a single example, the rise of China as an economic power in the 2000s has brought with it a very rapid increase in scientific productivity, both for that country and for the globe as a whole, as talented Chinese students and researchers have increased their written output, immigrated to other nations, and expanded the pool of global technical talent. China's economic and scientific rise, however, has depended on English as the medium of global contact and publication, and so has strengthened that language's importance in turn.

While English has expanded, thousands of native tongues have become endangered and extinct, leading some scholars to presume a causative relation. Generally, too, the worldwide reach that English now commands in the sciences has inevitably attracted a degree of resistance and anxiety, as it seems to work against sacred categories such as diversity. There is also concern that many native and national tongues will not be able to "keep up" with English in terms of scientific vocabulary and expression, and will thus "die" in this crucial domain. In reality, language threat is a far more multifarious phenomenon. Four factors—children, education, ideas of opportunity, and cities—appear to be the determinants in language change today. For technical professions with a strong international dimension, these factors are not just relevant. They are critical. As we will see in the following chapters, when the evidence is gathered, it becomes clear that although national science in national languages is not at all in danger of going extinct, no country with ambitions to advance in the scientific realm can hope to succeed without use of the English language—so far has the situation evolved. Use of English, in other words, means access to the great majority of the world's technical knowledge. It means an education system that can teach the language adequately, as a skill, thus providing for such access. It means an expanded arena of employment, domestic and international, for those possessing such skill. Finally, knowledge, education, and work in the sciences (as in all the professions), already concentrated in urban areas, will become even more so as cities take over the future of human society.

Whatever now has an impact on the global position of English has an impact on the sciences—and vice versa. It remains to be seen how this came about, where English in scientific practice stands today, and what this may mean for the future. Only when we know these things can we adequately answer the question of this book.

(3)

English and Science

The Current Landscape

The deepest sin against the human mind is to believe things without evidence.

T. H. HUXLEY

Andre Geim is a shy man with a thick Russian accent, dark wavy hair, and a warm expression, belied by a tendency to offer unexpected opinions. He received the 2010 Nobel Prize in Physics for his work on graphene, a unique substance of pure carbon, one atom thick. Dr. Geim has an interesting history. Born to German-Russian parents, both engineers, he spent his first seven years in the city of Sochi on the northeast Black Sea, a place of pebbly beaches and limpid sunsets that Stalin turned into his favorite resort. Since Joseph Stalin was not Russian but Georgian, therefore an immigrant, he tended to enjoy places far from the capitol.

In the late eighteenth century, Catherine the Great, a German herself (born in Prussia), invited Europeans to help settle the Volga River region, promising fertile land, exemption from taxes, and cultural autonomy. A century later, after the end of the Caucasian War (1817–64), the Russian Empire laid its hand over the whole North Caucasus region, executing or exiling the resisting Circassian natives, then inviting the adventuresome from central Europe once again to come settle the "new" territory. Geim's ancestors were possibly part of both groups of settlers, certainly of the former. During World War II, Stalin had many of these former Germans rounded up and sent to the gulag, Andre's father among them. He was allowed to return in 1949.

In 1965, the Geim family moved east to Nalchik, capital of Kabardino-Balkaria, one of the republics along the rugged northern border of Georgia that also include Chechnya. Andre attended a school that offered an intensive curriculum, including English, and he graduated with high marks. Despite the discrimination he suffered because of his name, he much wanted for university to attend the Moscow Engineering Physics Institute, Russia's top institution (it remains among the very best, with a curriculum in physics, mathematics, experimental studies, and English). High scores on the entrance exam didn't help get him in. Only later did he realize that as a "German," he was considered a flight risk to the West (he was first called a Russian when he was thirty-two years old, after he had left the country). He went instead to the Moscow Institute of Physics and Technology, earning his PhD in 1987, during the twilight Soviet years and soon after the Chernobyl disaster. Following collapse of the USSR, members of his family, including his father, immigrated to Germany.

Geim took a research position at Nottingham University in the United Kingdom. He says, "The people had the equipment and the time to do things, and it was extremely interesting; the facilities were incomparable. The Soviet Union had a very good theory school, but it did not have the facilities to really do experimental research. It was a different world. At that time I determined to get a position in the West and leave Russia." When his fellowship ended, he became, in his words, "a globetrotter," accepting posts at the Ørsted Institute in Copenhagen, the University of Bath, and again at Nottingham. In 1994, he accepted a position at Radboud University and stayed for six years, obtaining Dutch citizenship. (In all these positions, Geim taught his courses in English and published his articles, often with collaborators, in this language.) In 2000, he decided to leave the Netherlands. Offers arrived from around the world, but he chose the University of Manchester because of his experiences in the United Kingdom. Still, the Netherlands has not allowed Geim to escape entirely; in 2010, Radboud University appointed him professor of nanoscience and materials. He has also been involved with his former institute in Russia, working on a new type of graphene transistor.

Geim has often been invited to offer his opinions on various issues.

After giving the keynote lecture (in English) at the Second International Nanotechnology Conference in Tel Aviv, he was interviewed by an Israeli newspaper, which perhaps suspected his Jewish background. Asked about the brain drain, a heated issue in Israel, he said, "You won't like my answer. Science isn't football, and a scientist isn't a player on a team, but a worker for all of humanity. The brain drain shouldn't be stopped . . . free movement should be allowed."[1]

Key Historical Points

When did modern international scientific communication begin? And where did English fit into this? It seems safe to answer the first question by pointing to the beginning of the scientific journal as a distinct new form in the flow of knowledge. The journal, in fact, served three functions that, in combination, distinguished it from earlier types of exchange (for example, correspondence): first, its ability to disseminate new knowledge to many readers in many places simultaneously; second, its role as a repository, where knowledge could be stored and retrieved (again, by many, dispersed readers); and third, its role in establishing a professional rhetoric for science, in terms of both the "scientific paper" and its structure and style of discourse.

The first true scientific journals in the modern sense were not published in Latin. The *Philosophical Transactions of the Royal Society of London* and the *Journal des sçavans* both appeared in 1665, in English and in French, as the publications of their respective intellectual societies, the Royal Society and the Académie des Sciences. That they each chose the vernacular made perfect historical sense. England at this time acted as the center of the Scientific Revolution, with the Royal Society proclaiming the need to make scientific language simple and accessible by avoiding the use of complex, ornate, Latin-like speech. France, meanwhile, had become under Minister of Finances Jean-Baptiste Colbert the most powerful nation in Europe as well as a nexus for the new "experimental philosophy," with French soon the new lingua franca of eighteenth-century scholarship. These two journals were followed in 1682 by the *Acta eruditorum*, published in Leipzig and patterned after *Journal des sçavans*, but printed in Latin.

By today's standards, these early journals of science were diverse, even overly enthusiastic collections of material, varying from reports of single observations to long lectures, book reviews, essays on original experiments, maps, mathematical discussions, extracts from foreign literature, and more. During the 1700s, these periodicals multiplied considerably and remained publications of their respective scientific societies or their individual editors. Moreover, they were not at all shy about "borrowing," begging, and stealing material, often in the form of translations, from one another. As traced by David Kronick and others, they expanded most rapidly after 1750—by the end of the century, well over 400 were being produced, with the greatest number issuing from the German-speaking states (of which there were many at this time, before unification in the late 1800s).[2] Most consisted of secondary, derivative information and did not survive very long. Even the *Journal des sçavans* lasted little more than a century before being reformed into a literary publication. From the earliest period, only the *Philosophical Transactions of the Royal Society* continued uninterrupted into the twentieth century, down to the present. Its influence over time has been strong; even before 1700, a number of the most prestigious scientists on the Continent, including Leibnitz, Mercator, Cassini, and Huygens, had joined the Royal Society, and this trend only increased as a pattern over the next hundred years, when the ranks of members and contributors came to include the likes of Bernoulli, Maupertius, Buffon, Euler, Laplace, d'Alembert, and Benjamin Franklin.

With regard to the international exchange of purely original research, however, few journals provided a better model (and thus a window to later centuries) than *Observations sur la Physique, sur l'Histoire Naturelle et sur les Arts*, a monthly published by François Rozier, beginning in 1773.[3] Rozier fully intended to do something new, to focus on new findings throughout Europe and to appeal to those actively pursuing scientific work. Among the luminaries whose work appeared in his pages were Linnaeus, Priestley, Scheele, Lavoisier, Berthollet, and Black. The first issue of Rozier's journal (as it is known) carried an introduction in which the Abbé had some interesting things to say about science and language at the time.

> Scholars have always been aware of the infinite advantages that result for the advancement of science from a lively and extensive commerce which joins all the members of the Republic of Letters through the regular communication of ideas, views, endeavors, and projects. . . . However, the majority of [current] academic collections

> are published in their national language and are printed several years after [they] have been read.* During this time one remains ignorant of facts which could be of the greatest utility for the sciences . . . scholars of two different nations work a long time on the same problem. . . . These then are the reasons which persuaded us to undertake this collection and we present it with as much confidence to the scholars abroad as if it were their own publication. It is written in a language which is today that of all those in Europe who have received any education.[4]

Rozier was not here speaking of Latin, of course, but of French. No less significant are the editor's justifications for his journal—the need for any "Republic of Letters" (the word *scientist* was invented only in 1833 by British philosopher William Whewell) to create a fully international, evolving, up-to-date repository of "ideas, views, endeavors, and projects" in a single language shared by all in this "Republic." Science, in other words, is cast as a democracy, whose citizenship requires that one be both an active researcher *and* a speaker of the communal tongue.

During the nineteenth century, this sensibility weakened, as nationalist science became more the norm, reflecting conflict among the European powers. The number of journals grew hugely in England, France, the German-speaking states, Holland, Scandinavia, and the United States, from hundreds to thousands overall, with a great many focused on individual fields and subfields, which developed as science matured into a major profession. German certainly did rise as a new lingua franca in the late 1800s, yet its purview mainly covered portions of the physical sciences, particularly physics and chemistry, plus mathematics and medicine. French remained essential, and English kept a smaller but still important international position, and not only because of Britain's colonial possessions. England had been the forge of both the Scientific and the Industrial Revolutions, a fact that demanded some accommodation to this tongue by many researchers and engineers (as well as translators). Britain was also the birthplace of arguably the two most influential scientists before Einstein—Isaac Newton and Charles Darwin. Moreover, English was

* It was common practice for new experiments and findings to be first read before one or more scientific societies before appearing in print.

the language of the United States and Canada, creating a transoceanic dimension. While the United States couldn't be called a scientific power until well after 1900, it did serve the purpose of cultivating international prestige for English in two ways: first, by its key role in the so-called second industrial revolution of the late nineteenth and early twentieth centuries; and second, by becoming the major technological influence on Japan, the most advanced nation in East Asia.

Studies that have traced the development of scientific journals and their languages after 1900 agree on two major patterns. First, the number of journals grew significantly but moderately up to the end of World War II, then exploded into a period of unprecedented growth that continued at least into the 1990s.[5] Second, German, French, and English were the dominant languages of these journals in roughly equal proportions until after World War I, at which point German and French began to decline—German quite rapidly and French reaching a plateau between the 1920s and '40s, then falling once more (see figure 3.1 below). The quick falloff for German, which had been so recently and for a number of decades a crucial lingua franca, was largely due to the damaged image of the country and its scientists, some of whom were known to have participated in the war effort. During the early 1920s, a boycott was staged against German scientists and the German language by scientific academies in countries that had fought on the Allied side.[6] The boycott, which did not last long but ended less than a decade before Adolf Hitler's rise to power, applied to international meetings and symposia. It thus had an ostracizing effect. Though much intellectual commerce continued with German scientists, particularly in physics, the country had lost its premier standing among Western scientific powers. Nazism and World War II would soon complete Germany's fall from grace, with most of its prestigious researchers leaving the country and emigrating to the Soviet Union, England, and above all, the United States.

It was at this point, following the war, that English began its final ascent in the sciences, propelled by the unrivaled position of US science. This is a reality that has only rarely been given its full due. American science gained a breathtaking advantage, historically speaking, from the war and its devastations. Europe, long the core of modern science, lay in ruins and faced reconstruction for nearly a decade, with its eastern countries, including half of Germany, occupied by the Soviet Union. Itself greatly weakened, the Soviet state could muster the technical might to build its own nuclear weapons, but remained unable to erect a full-scale scientific culture for many years, a condition exacerbated by Stalinist ideology and its constraints on research. In China, civil conflict followed by the Great Leap Forward and the Cultural Revolution kept scientific work minimal beyond the 1960s. Japan saw its cities in ashes and its islands occupied, even as Korea fell into a horrific war of its own that tore the country in half.

Among the world's major nations, the United States stood alone for all practical purposes. In the aftermath of World War II, its cities and universities were undamaged, its infrastructure hugely expanded, its resources enormous, and its research capabilities in many areas powerfully advanced. No country or even group of countries could begin to compete with the massive scale of the US scientific establishment, which the war had raised to great heights of productivity and (no less important) status in the public and the political eye by its contributions to ultimate victory, most notably in the form of the Manhattan Project, which produced the first US nuclear weapon. Project organizer Vannevar Bush's vision of a healthy US superstate, powered and defended by scientific advancement fully underwritten by the federal government and providing new knowledge to the rest of the world, fell on open and willing ears in the late 1940s. With its enormous wealth, the United States rapidly enlarged its scientific culture further, creating the federal-grant university system and a sprawling network of government labs and using its powers of procurement (especially for military and defense purposes) to stimulate corporate research and development (R&D) as well. The war ensured that it would take decades for scientific powers such as France, the Netherlands, and Germany to fully regain the range and status they formerly had. By that time, English had become the dominant language of international science.

Such a brief and brutal summary does little justice to the detailed story of science and English. But it does underline a few crucial points. From early on, after the collapse of Latin, the value of a lingua franca for science was recognized. Indeed, it was understood to be commensurate with an international community at its most collegial and productive. Nationalist science grew up around this view—for most of the modern era, a tension has existed between national and international allegiances in science, between loyalty to country and to discipline. This tension broke into the open with the exile of German scientists from many international meetings after World War I. A fair number of the most prestigious of these individuals—Max Planck, Wilhelm Roentgen, Ernst Haeckel, and Paul Ehrlich, among others—had signed the infamous Manifesto of the Ninety-Three, a declaration publicly supporting the German government and military in their aggression against France and England. A code of honor had been broken: instead of working for the advancement of knowledge and the progress of humanity, these men had chosen national conquest, war, even (conceivably) the killing of fellow researchers. Rozier's "Republic of Letters" and its higher ideals had been betrayed. That the scientists involved spoke what was then a functioning lingua franca made the situation all the more charged.

Meanwhile, American advantage in the sciences after World War II was needed for the final ascent of English as a global lingua franca. Yet the full truth of this ascent is more complex. US scientific power accelerated a trend that was already in motion, a trend that England had initiated, that the declining status of German science helped promote, and that, by the 1930s, England together with the United States thrust into full advancement. Data on factors including scientific publications and foreign-language preference in education bear this out, showing an increase in the use of English at the expense of other tongues well under way by the end of the 1930s. What seems clear today—with publications in English from European nations outnumbering those from the United States, and papers from China, Japan, and Korea together approaching this level, too—is that America finds itself no longer a required participant for the prestige and spread of its native language in science. A global tongue, as we have noted before, is a language at once denationalized and supranational. It is, if we accept Rozier's ambition as a goal, required for that "commerce which joins all the members of the Republic of Letters through the regular communication of ideas, views, endeavors, and projects."

Circulatory System

Workers in the sciences have long been merchants of knowledge. Today, as Andre Geim would say, their trade routes lead everywhere. Science in the twenty-first century has been globalizing at a rate barely conceivable a few decades ago, when its future seemed firmly in the grip of a dozen or so wealthy nations.

How do we know this? The story of Andre Geim is emblematic. Researchers now migrate between continents and hemispheres—between Asia and Europe, Asia and North America, Latin America and North America, Africa and Europe. Increased mobility reflects the rise of scientific capacity-building in nations throughout the world, including digital technology, which has not only accelerated the flow of knowledge but allowed researchers to create international networks more readily than at any time in the past. If we survey the changes now ongoing in what can be called the "support system" for science, we come face to face with several large-scale factors. Two such factors are the rise in R&D funding and in the number of researchers, particularly in developing countries.

Are there quantitative measures we can examine? Indeed there are. Organizations such as the United Nations Educational, Scientific and Cultural Organisation (UNESCO), the European Commission, and the National Science Board in the United States collect such data routinely. The information shows that R&D spending worldwide entered a surge period in the 2000s: in the ten years between 1999 and 2009, such spending more than doubled, from $641 billion to $1.3 trillion. The greatest increase occurred in East Asia, which grew its total more than threefold, surpassing Europe in 2002 and approaching US levels by 2009 ($400 billion).[7] These big numbers, however, can obscure another essential fact. In the brief five years between 2002 and 2007, developing countries increased their combined R&D investment more than 100%, and during the next two years they increased it another 50%. It is clear, in other words, that science and engineering have become primary foci for nations everywhere that seek economic advancement. In 1990, North America, Europe, and Japan accounted for more than 95% of global R&D spending; in other words, all developing nations together didn't even account for 5% of global R&D. By 2009, China alone represented 13%, India 2.5%, and Brazil 2%.[8]

An even better metric, perhaps, is people. Looking at the 1996–2007 time period, we find that the estimated number of researchers in wealthy, Organisation for Economic Co-operation and Development (OECD) countries increased by 9%, a much higher rate of growth than for the general population, which averaged less than 2% for these nations. The exception was the United States, whose total research labor pool grew by only 0.6%

(England's grew by 2.9%). In developing nations, meanwhile, researchers expanded their ranks by a remarkable 56%—in a single decade, these nations increased by more than half their total number of working scientists. If we break the data down further, the figures for various nations and regions are these: China—76% increase; India—34%; Brazil—74%; Argentina—48%; Turkey—108%; Latin America and the Caribbean—48%; sub-Saharan Africa—33%. From contributing less than 15% of the world share of researchers in 1990, developing countries were employing nearly 40% by 2007.[9]

These numbers are stunning. They are all the more striking because in many of the nations involved, high-level economic support for R&D is a new thing and, given political volatilities, not at all guaranteed for the long term. Yet these nations are investing in and encouraging their youth as if future priority for research is secure. Certainly, many of them remain at an early stage of scientific capacity-building. It will take Vietnam or Peru decades to achieve a level of research rivaling that in Denmark or New Zealand. Yet they are now under way to achieving such a goal. The first half of the twenty-first century appears destined to be the era when the greater part of the developing world joins the rise of modern science in earnest.

Another key reality, however, is that the scientists and engineers they train may not be theirs for very long, or only intermittently. "Free movement," in Andre Geim's terminology, is still far from a universal reality, but it has nonetheless become a global one on a scale never before seen. The numerous reasons why today's scientists decide to change jobs and locations—whether for better employment, funding opportunities, more sophisticated facilities, short-term projects, teaching assignments, multiyear collaborations, advanced training, or family considerations, or because of ethnic conflict, persecution, political oppression, or war—are very much part of the world fabric as it exists, including its many inequalities. Researchers, in other words, are caught up in the vicissitudes of globalization, too, though increasingly as the carriers of skill, talent, and economic potential.

These vicissitudes have many dimensions, some brilliant, some less so. The global reach and power of large R&D corporations, encompassing expertise from many fields, can lead to tremendous productivity and fascinating arrogance. Geim himself provides an example. He recites an interaction he had at a conference where he asked a representative from a large multinational electronics firm if it might be willing to help sponsor a patent on graphene that Geim and his team were thinking of filing (it is expensive to keep a patent alive over time). "We are looking at graphene," the man replied, "and it might have a future in the long term. If after ten years we find it's really as good as it promises, we will put a hundred patent lawyers on it to write a hundred patents a day, and you will spend the rest of your life, and the gross domestic product of your little island [Britain], suing us."[10]

Brain Drain or Gain?

Geim's own history, meanwhile, suggests that the term *circulation* might be better applied to the migration phenomenon than *brain drain*. But this will inevitably depend on one's perspective. Observations of scientific migration demonstrate the phenomenon's actual complexity. It remains the case that the flow of talent moves mainly from from poorer to richer, and also among the rich. But this generality hides far more interesting and suggestive realities.

> South Africa, Russia, Ukraine, Malaysia and Jordan have also become attractive destinations for the highly skilled. The diaspora that has settled in South Africa originated from Zimbabwe, Botswana, Namibia and Lesotho; in Russia, from Kazakhstan, Ukraine and Belarus; in Ukraine, from Brunei Darussalam; in the former Czechoslovakia from Iran; in Malaysia from China and India; in Romania from Moldova; in Jordan from the Palestinian Autonomous Territories.[11]

The idea of brain drain as a zero-sum, one-way transfer of human scientific capital enfeebling one country while fortifying another is out of date. In fact, it was always something of a mercantilist idea.† In many

† Mercantilism was a collection of economic policies practiced by European states in the sixteenth through eighteenth centuries, based in part on the idea that wealth depended on a nation's exporting more than it imported. Adam Smith's *An Inquiry into the Nature and Causes of the Wealth of Nations* (1776) is considered to have dealt a death blow to mercantilism as a philosophy of trade.

cases, it simply wasn't true at the level of economic return. Economists now know, for example, that the money sent home by skilled emigrants in the past more than paid for the education they received on the government's dollar.[12] By not returning home, these emigrants also avoided the fate of reentering an environment unready for what they had to offer. This leads to a key point.

The term *brain drain* was coined by the Royal Society in the early 1960s to exhibit alarm over the loss of cerebral talent, scientific talent in particular, which in the wake of Word War II seemed to be flowing in great rivers from Europe to the United States and Canada. But rather than lay waste to European science, the migration led to new policies in favor of home science. It was evident that levels had to be raised: huge disparities in rewards, advancement, and the very ability to practice the best science had created the pressure to migrate in the first place. Europe therefore came to support more funding for training, improved labs, higher salaries, better opportunities. Within a single generation, the drain turned into a massive gain, helping to modernize a system still rooted in the nineteenth century. But no less did it aid science in another way. It accelerated the scale and international flow of knowledge during the Cold War era, setting the stage for the high levels of networking and collaboration we see today. By the mid-1990s, the output of European science, as measured by peer-reviewed articles, surpassed that of the United States.[13]

Now the tale is being retold for the developing world. Twenty years ago, scientists and students who left their home countries for work and training in the West had little reason to return. Disparities in every aspect of scientific culture guaranteed migration of talent. Even the requirement by some countries that those trained on state funds had to return and work for a specified time period did little to slow migration. Nationalism in these places wasn't any more a routine motive for researchers than it was in Europe. Between 1978 and 2006, over 1.06 million Chinese went abroad to study, the great majority of them in science and engineering; 70% did not return.[14]

Since 2000, the situation has changed. Developing countries have learned the lesson that Europe learned in the 1950s. Nations including China, India, Korea, Brazil, and Taiwan have been building up their own research infrastructure so that they can better keep or lure back native talent. Such helps explain a fair portion of the increases in R&D spending for these nations noted above. In all these countries, returning researchers have been key agents in determining what needs to be done, in order that labs and facilities, the quality of journals, and other aspects of science are put on a par with those of developed nations. As scientific culture in places such as China and Brazil has advanced, more researchers have remained or returned home—at least for a while.

Because now there are other opportunities. Scientists with strong reputations can do what Andre Geim did—become "globetrotters," accepting positions in more than one country and institution, even at the same time. Their opportunities come both from their own careers and from the changing nature of science itself—its greater reliance on digital technology and related communicational possibilities, the interdisciplinary nature of much new research, and the emergence of international collaboration and the networks on which it is based as central parts of the "invisible college."[15] Global mobility in terms of people has grown as much from self-organizing networks as from cheaper transportation, intercultural contacts, and rising incomes in much of the developing world. The old two-way street of drain-and-return has become a great, multibranched highway system spreading outward in many directions. By the mid-2000s, it was evident that "in the context of increasingly knowledge-based economies . . . the current situation tends to imply many types of return [for scientists], circulation, and recirculation."[16]

There has also been much interest in drawing scientists from English-speaking countries to come and teach in non-anglophone countries. Another type of effort seeks to create collaborative programs with European and American universities. Scientists with strong publication records are being offered dual positions, in their native country and in the country where they were trained (and did postdoctoral work), with the pairing of China or India and the United States or Britain being the most common.[17]

Still other factors need to be considered in the new circulation. One of these is the rise of multinational firms in developing countries, which provides a brand-new type of attraction for native scientists. Knowledge- and technology-intensive industries, from computer chip makers to pharmaceutical companies, generally have become a major segment of the global economy. By 2010, they were generating over 30% of world economic output, a formidable $18.2 trillion.[18] Key here is that multinationals have been decentralizing their operations, including research, resettling in both developing and developed nations to gain better access to local markets and talent.[19] Scientists who return to their native countries might therefore end up working somewhere else—a Korean nuclear researcher helping to guide reactor construction in the United Arab Emirates or a Chinese agronomist sent to evaluate cereal harvests in Mozambique.

These truths are brought home by a mining geologist who, when asked about these trends in 2010, told me,

> At the end of the '90s, we all felt the industry was in trouble. Base metal prices were down; people being let go. Within five years, everything changed. China and other emerging countries sent mining into a boom. Today there's exploration all over the world. China now uses more metals than Europe, Japan, and the U.S. *combined*. It's an amazing time. There are some chaotic ups and downs, of course, but the amount of work for scientists, in the lab and the field, is way up. . . . I'm used to relocating—nobody in this business writes their address down in pen—but everything's global now. Since 2005, I've worked with people from Japan, India, Sweden, Russia, Australia. The only way this works is that everyone speaks fairly good English. Gone are the days when an American or Australian company would hire only Westerners. That approach makes no sense anymore.

For scientists from developing countries, especially those who have studied or worked in anglophone nations, there are reasons to return home now. But coming home may be the first step in a journey that has only begun. Top researchers from any nation will find many possible routes open to them. A growing number judged to be at a high level, including Andre Geim, are finding opportunities in several countries on different continents. Such opportunities are not independent of language. As Geim's own career shows, global science relies on proficiency in English. The "free movement" he talks about is itself tied to such ability—just as (he might well say) the wandering scholars of the late Middle Ages were themselves bound to Latin. Our Korean nuclear physicist and Chinese agronomist will also have been chosen because of such ability. For such researchers, the world is not shrinking. By virtue of English and globalizing science, it is growing larger and more diverse with every passing decade.

Publications: Global Trends

Since the 1970s, peer-reviewed research papers, letters, and communications in major journals have been used as the dominant metric of scientific "output." Though now widely accepted by institutions, this approach will always be legitimate yet limiting, attractive yet incomplete. There are two main reasons for this. First, focusing on the journal literature leaves out many types of publications dealing with frontier research: books and monographs; online scientific databases; preprint archives; government reports and white papers; and publications from nongovernmental organizations, companies, international organizations (the UN, World Health Organization, etc.), and advisory bodies (National Academies), to name only some. Second, the tracking of journal literature has understandably relied on a very small number of sources—for decades, the Institute for Scientific Information, creator of the Science Citation Index (SCI), now part of the online citation index Web of Science (available from Thomson Reuters), the world's largest bibliographic database, was the single dominant supplier of relevant data. In the 2000s there arrived competitors, Scopus (Elsevier) and Google Scholar. All three services are expanding their selection of material every year, but they are nonetheless limited by certain criteria. For the journals they choose to include, SCI and Scopus require signs of professionalism, such as regularity of publication and quality of editorial standards. But they also demand that articles be in English, have abstracts in English, or at least have their references in English. This is most true of SCI (through Web of Science), which has selected comparatively few non-English sources over the years. Indeed, in describing its own criteria, SCI is very straightforward about this.

> English is the universal language of science at this time in history. . . . There are many journals covered in *Web of Science* that publish only bibliographic information in English with full text in another language. However, going forward, it is clear that the journals most important to the international research community will publish full text in English. This is especially true in the natural sciences. In addition, all journals must have cited references in the Roman alphabet.[20]

Large nations, such as China and Brazil, produce many thousands of scientific periodicals in non-English languages. This applies on a regional level, too. In 2009, for example, at least 6,555 scientific research journals were published in Spanish and Portuguese in thirty-two countries spread across Latin America, the Caribbean, Spain, and Portugal, most covered by the Spanish-language database Latindex; only a small fraction are included in the Scopus and Web of Science indices.[21] While most of these publication may well be of local interest (as are many periodicals produced in the United States and Europe), it seems a legitimate criticism that more of them likely deserve inclusion in databases claiming to offer "comprehensive" services to the research community.[22] Both Web of Science and Scopus have made efforts to respond to such criticisms by expanding their non-English selection of journals. Early comparisons suggest that Scopus has been more receptive to such journals, and that Google Scholar is also able to retrieve more non-English citations than SCI.[23]

Yet there remains another factor. The ultimate power these databases—above all SCI—now wield on the international stage is considerable and works very much in favor of English. In particular, SCI has cast itself as not only the arbiter of scientific "output" but also the provider of indicators for research "quality," based on citation numbers and related indices. The most widely recognized and influential of these indicators is the "impact factor," defined on the basis of citation frequency. Since the 1990s, the use of this and related SCI indicators, and bibliometrics in general, has been widely embraced as a way to evaluate individuals, university departments, entire institutions, and research projects. As such, related indicators have become factors in hiring decisions, and granting of tenure and funding, too. Any uses of this kind will continue to drive the move toward publication in English.[24]

Peer-reviewed journal literature in English mainly reflects academic research of the highest global status. It also has the greatest visibility among scientists and the media. It thus remains the most read and cited venue, and is thus where scientists most want to see their work appear. If not a wholly accurate measure of "output" (the corporate sector produces an enormous amount of proprietary science), it is a gauge of what might be called competitive presence in cross-country comparisons.

At this level, certain trends emerge that parallel those noted earlier for R&D spending. The most widely discussed pattern in recent years has been the astonishing rise in scientific papers from China. Between 2002 and 2008, the total number written by Chinese researchers rose by no less than 174% (from 38,206 papers to 104,968), based on Web of Science data. No small part of this growth has apparently been due to the changeover in many Chinese journals to English publication. Such spectacular growth, matching as well the rise in researcher numbers and in R&D spending (scientific capacity-building), is then posed against a "mere" 20% increase in written output by the United States over the same time period (from 226,894 to 272,879) and a 24% rise from the European Union (from 290,184 to 359,991).[25]

In truth, the upsurge from China, notable though it may be, doesn't make the increases from developed countries trivial by any means. Recall that the labor pool of researchers in the United States grew very little between the late 1990s and late 2000s; a 20% rise in scientific papers therefore means a major rise in productivity (papers per researcher). This also holds true for Europe. In purely numerical and—for scientists struggling for tenure or promotion in such a brutally competitive environment—merciless terms, the gains made in these countries with mature scientific cultures shouldn't be less impressive than those by the Chinese, who, after all, are new on the global scene and begin from a far smaller base. The publication engines in the United States and Europe have hardly slowed. As scientists themselves will confirm, those wheels spin more quickly than ever.

Yet the real story lies elsewhere. First, the growth from China, like that from Western countries, is a global success story, not merely a national one. Again, this growth is gauged for papers published in English, not Chinese, thus in a tongue accessible to the greatest number of scientists worldwide. Chinese researchers are not seeking to weaken or counter the dominance of English, or offer an alternative to it. On the contrary, they are greatly advancing it. Global scientific success for China means success in English. This is not an irony; it is an inevitable consequence of a global scientific language. Given recent trends, the Chinese could even match US levels in peer-reviewed English-language "output" by about 2025, perhaps sooner. But what would this mean? However interpreted in political terms, its fortifying impact on scientific English would be beyond question. Aside from America, China may well be the most powerful force behind the spread of English in science.

There is another chapter to the story. Figure 3.1 reveals that major

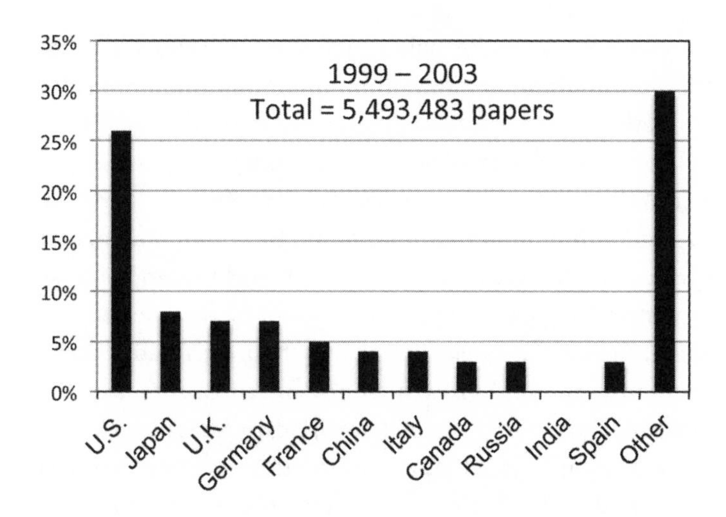

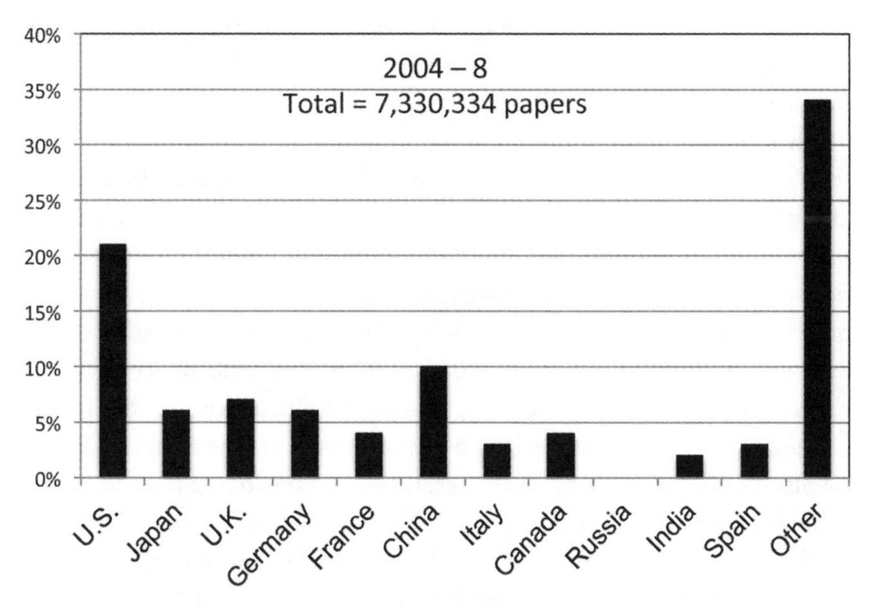

FIGURE 3.1 Percentage of global scientific publication for various countries, given for two five-year periods: 1999–2003 and 2004–8 (inclusive). Graph size correlates with total number of papers. Data from Royal Society of London, *Knowledge, Networks, and Nations* (2011).

changes took place in global scientific publication between 1999 and 2008. In the first half of this period, between 1999 and 2003 (upper graph), Europe, Japan, and the United States still accounted for two-thirds of the measured global "output" of peer-reviewed papers; by 2004–8 this had fallen to just over half. Another way to look at this, of course, is to flip the lens: developing and emerging countries (shown by Other, China, and India bars), representing the great majority of the world's nations, rose from 34% of all "output" in 1999–2003 to nearly half by the later 2000s. Indeed, by 2004–8, the largest category by far in figure 3.1 had become Other. This included very rapid growth from Eastern Europe and the Baltic states (66% over 1999–2003), Brazil (111%), North Africa (73%), Iran (418%), Turkey (107%), South Korea (92%), and portions of Southeast Asia (more than 40% for Malaysia, Thailand, and Singapore especially). Latin America, though its increase may not have accelerated so quickly—much of the region has only recently emerged from dictatorship—is nonetheless on an exponential growth curve, too.[26]

These data tell a tale of tremendous historical importance. Rather than the decline and fall of US or European science, they reflect the remarkable growth in research and the level of international participation that is happening everywhere else—the globalization of science itself. China is not at all alone; more of the world, starting from a low base, is now seriously beginning to add to the recognized global stock and flow of scientific knowledge. We see no empires crumbling; this is no tale of armies at the gates. Very much to the contrary, it is now a global fact—and a highly beneficial one, if you happen to be a scientist—that many nations barely involved in modern research a generation ago have joined the generators of new knowledge. Another part of the story given in figure 3.1 is that total "output" grew profoundly during the 2000s: more of the world is producing more science of recognized quality than ever before.

Collaboration Trends: Hands across Borders

Publication "output," tracked by nation, offers a competitive view of science. This is the institutional view, anxious about priority, rank, economic impact. We might therefore ask: what part of this "output" is cooperative, with coauthors from different countries collaborating? The answer: a goodly and ever-growing amount. Figure 3.2 suggests that in-

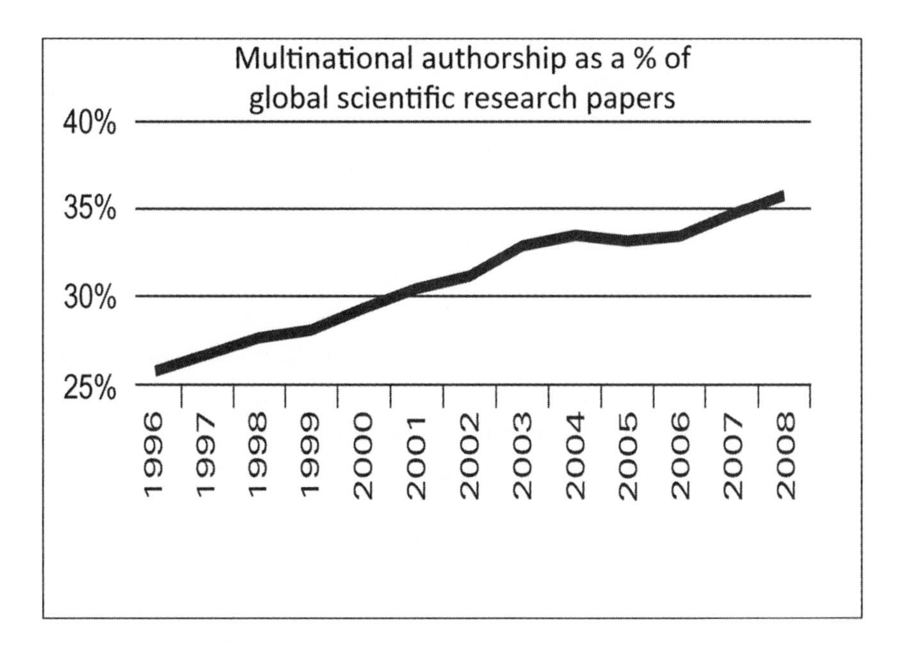

FIGURE 3.2 Graph showing growth in multinational authorship as a percentage of the world's scientific papers (as tracked by the Scopus database) over the time period 1996–2008. Adapted from Royal Society of London, *Knowledge, Networks, and Nations* (2011).

ternational collaboration now stands as a primary dimension of scientific work worldwide, having advanced between 1996 and 2008 from roughly a fourth to over a third of all peer-reviewed articles in journals tracked by Scopus.[27]

For scientists themselves, this will sound correct. The number and scale of international research endeavors have grown vastly, even to legendary proportions, defying all previous limits. If, in the early 2000s, the Human Genome Project yielded research papers with hundreds of authors from more than a dozen countries, this was grandly surpassed in March of 2010, when a paper analyzing data from the Large Hadron Collider (located along the French-Swiss border near Geneva) was published in *Physics Letters B* carrying no fewer than 3,222 coauthors from thirty-two nations.[28] So-called big science has never been bigger or more international. But the emergence of multinational collaboration can be found everywhere in science today. There are other big-time examples: the International Space Station, the Hubble Telescope, the International Thermonuclear Experimental Reactor fusion effort, the Intergovernmental Panel on Climate Change, the Global Biodiversity Assessment, the Long Term Ecological Research Network, the International Geological Correlation Programme, and so on. But there are also innumerable smaller projects involving subjects like the study of isotopes in lithium deposits of the Salar de Uyuni, Bolivia, or the variable effectiveness of the tuberculosis vaccine bacillus Calmette-Guérin in an immigrant population. In nearly every field, multinational research is now routine, common, or becoming so.

Thus, we should expect growing levels of coauthorship. This is indeed what we see. A closer look at the data offers some interesting details. No nation grew its degree of collaboration during the 1996–2008 period more than the United States, whose numbers nearly doubled, from around 17% to 32%.[29] At the same time, no other advanced nation was at such a low level, proportionally speaking, in 1996. Other publication hubs, including England, the Netherlands, Germany, and France, rose from an average of about 30% coauthorship in 1996 to 45% or higher by 2008. For instance, Andre Geim while in England has seen the number of his coauthors rise from two or three to a dozen during this same time period, with contributing researchers hailing from Russia, Spain, the United States, Greece, the Czech Republic, Portugal, India, France, and the Netherlands. For western European countries, collaboration has been a way of scientific life for many years due to geographic proximity, historical connections (including scientific ones), and regional programs that share funds, expertise, and facilities such as the European Research Area (launched in 2000). But the recent surge is all the more notable for this: rather than having reached any sort of peak after many years, in roughly a single decade collaboration grew from a third of all publications to nearly half. Moreover, the data show that since the late 1990s, copublication with authors from outside Europe has grown most rapidly of all, suggesting that this collaborative tradition has itself expanded as part of the intellectual side to globalization.[30]

The next question we might ask concerns the "who." Coauthor networks have been found to be largely self-organizing; they aren't determined directly by R&D policies, corporate priorities, or other institutional influences.[31] Scientists decide whom they want to work with and why. Though governments and companies surely provide impact through funding, researchers are the ones who know the landscape of capability among their colleagues, who often have established networks already at their fingertips or know where to look to build them. Overall, coauthorship between scientists from wealthy and developing countries more than doubled during the first decade of the 2000s.[32] It is interesting, therefore, that scientists in the United States, though they coauthor papers internationally only about a third of the time, do so with investigators from the largest number of nations—as many as 173 by the year 2003.[33] We can read this in two ways: (1) American scientists are open and eager to collaborate with other workers everywhere; (2) scientists around the world often choose to work with colleagues in the United States. These two readings are not at all exclusive of each other. What they imply, however, is the centrality of English in global collaboration.

English in Publications: What Do the Data Say?

With the above as background, it becomes important, finally, to look at language data on natural-science publications. Since the late 1990s, a fair number of studies have been made on this topic tracing language use over time, both for science as a whole and for individual fields.[34] Overall, the data show very similar and consistent patterns, no matter the domain examined.

The graph of figure 3.3, using SCI data and covering a 120-year span, comes with the uncertainties already mentioned, yet displays the key patterns clearly. It shows percentages of global science publications produced in five major languages, as tracked in journals indexed by the SCI database and in American, German, French, and Russian bibliographies.[35] The trends confirm our historical discussion above. In the late nineteenth century, English, German, and French were all major tongues in science, with German rising rapidly, then falling in the 1920s. English at this stage began to grow in importance, finally surpassing German in the 1930s and becoming unrivaled in the post–World War II period. Yet this still meant that in the 1960s, as much as 40% or more of the literature continued to be published in German, French, and now Russian, with growing volumes in Japanese. It wasn't until the 1980s that English had come to dominate more than two-thirds of all publications. By the 1990s, the figure had risen to above 85% and by the 2000s even higher. If we use Dr. Geim as an example once again, we see that more than 150 peer-reviewed papers on which he has been coauthor, published since the early 1990s, have been published in English. Within a single generation, therefore, between 1975 and 2000, the most prestigious periodicals in just about every field switched over to this language, if they hadn't done so already. At the turn of the second decade of the twenty-first century, the international literature showed every sign of continuing the trend.

Interestingly, these basic patterns are echoed by the growing preference for English as the main foreign language in primary and secondary education worldwide. Figure 3.4, adapted from work by Yun-Kyung Cha and Seung-Hwan Ham, shows little or no preference before the twentieth century, growth into the early 1900s, then a rapid rise to full dominance between the 1940s and 1980s. By the 2000s, the level of favor granted

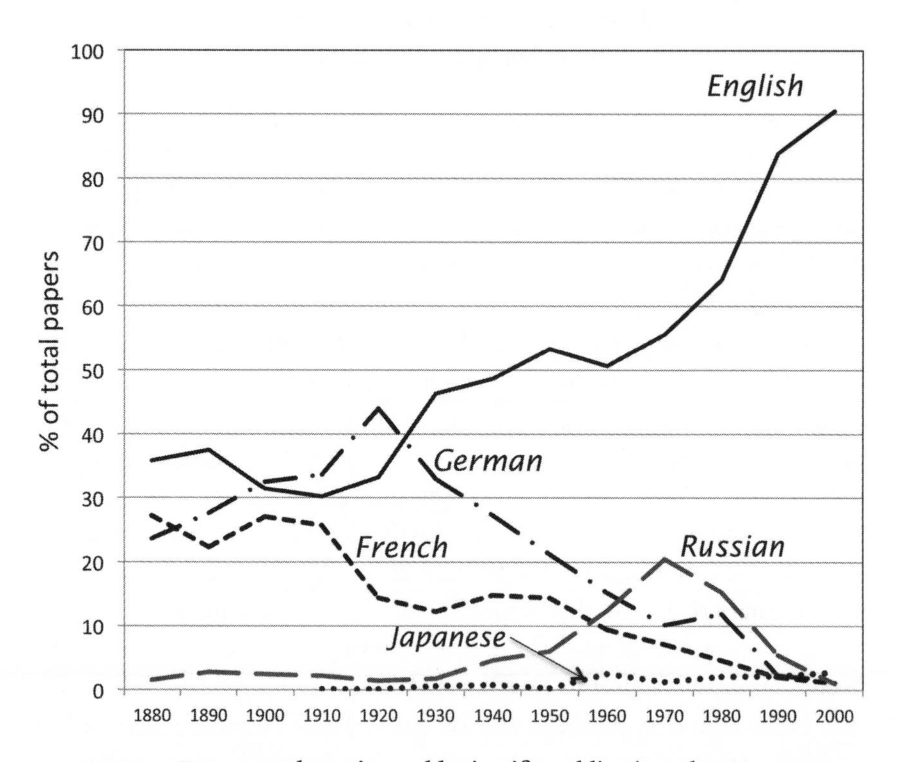

FIGURE 3.3 Language shares in world scientific publication, shown as percentages of total articles tracked for selected tongues over a 120-year period. Data have been adopted from several sources (see note 35).

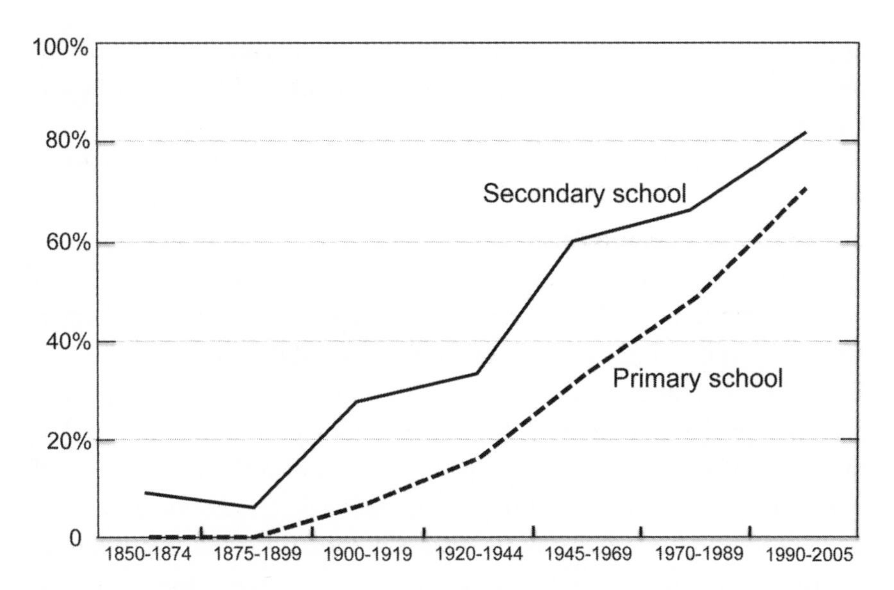

FIGURE 3.4 Percentage of surveyed countries with English as the dominant foreign language in public schools, 1850–2005. Adapted from Cha and Ham, "The Impact of English on the School Curriculum" (2008).

to English reached 70% for primary school and over 80% for secondary school, with the gap closing. These higher numbers, it should be mentioned, correspond to data from over 150 nations.[36] It is worth remembering, too, that the period of the late 1990s and 2000s also defines the time when developing nations in many parts of the world decided on policies to advance science and technology (and thus related training) for the sake of economic growth and social progress.

Case Study: The Life History of an International Journal

Since 1910, when it began publication, *Geologische Rundschau* has been among the most international of journals in the geological sciences. It has published papers on a wider array of subject matter than almost any other periodical in its class, thus attracting authors from many disciplines. By the late 1930s, it had become one of the world's top geoscience journals, a status briefly interrupted by World War II but then rejuvenated in the

1950s. Today, its official name as of 1997, *International Journal of Earth Sciences* (*IJES*), proclaims all these facts in unequivocal fashion. When it first appeared, and for most of its history, *IJES* accepted articles in German, French, and English. A few years after the launch of Sputnik (1957), it expanded this repertoire by providing abstracts for every paper in all these tongues, plus Russian.

We therefore have an excellent test case for language trends in natural science for the past one hundred years. *IJES* began publication when German was the dominant international tongue of modern science. What happened thereafter? Until 1930, the journal published only papers in German, probably reflecting the fact that all submissions were in that language. The first non-German paper appeared in the 1949 issue and was written in French ("De Jura," by D. Aubert). Also at this time, *IJES* began offering a summary of its total contents in French and English; the English version was only a page in length, while the French extended to four pages. The first article in English came the next year, 1950, in a special issue devoted to the geology of Africa (a large colonial presence by England), and was written by a well-known Cambridge University geologist, E. S. W. Simpson. It is at this point that the graph of figure 3.5 begins, showing the number of articles published in each of the three languages up through the year 2000.

From 1950 until the late 1960s, the number of articles produced in all three languages increased, because the journal expanded in size, as shown by figure 3.5. Taking 1965 as a sample year, *IJES* published 40 papers in German, 17 in French, and 9 in English, more than three times the average number of total yearly papers a few decades earlier. By the early 1970s, numbers for German and English papers were roughly equal, with French far lower. By the late 1970s, English had become dominant in the journal.

During this decade, too, another crucial pattern begins that helps explain the trend toward English dominance. For the first time, articles written in English by non-anglophone geoscientists (and having no collaboration with a mother-tongue speaker), including those from Germany, become common. In 1970, a total of 5 out of 59 papers (8.5%) are of this kind; by 1975, the number had more than doubled, to 13 out of 54 papers (23%); and by 1980, more than tripled, to 16 papers out of 54 for that year

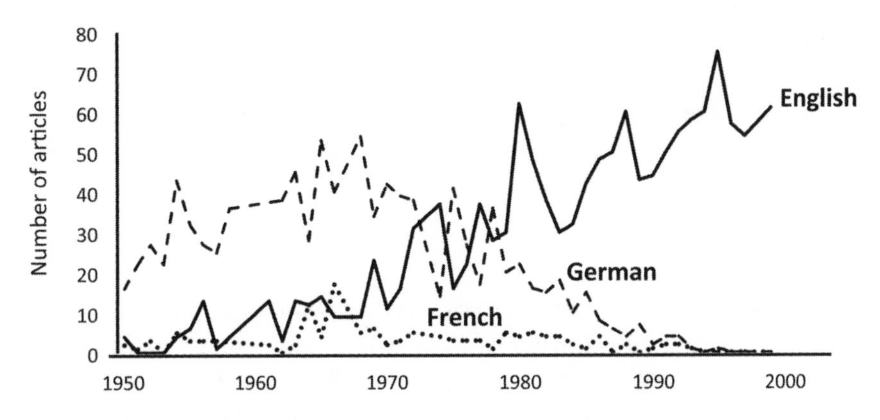

FIGURE 3.5 Number of articles published per year in German, French, and English in the journal *Geologische Rundschau*, 1952–2000. The journal changed its name to the *International Journal of Earth Sciences* in 1997.

(30%). Moreover, by 1980 these papers in English by foreign speakers had come to constitute over half of all papers in English, suggesting that the language clearly had been recognized by then as a budding global tongue for the geosciences.

After 1980, as figure 3.5 makes plain, the shift toward English is assertive, with German and French declining rapidly. Last vestiges of these languages are seen in issues the journal published in the 1990s. An address to the readership from the editorial board in 1993 was a presage of things to come: "Because the *Geologische Rundschau* is an 'International Journal of Earth Sciences,' the abstract will be written only in English, which is the dominant language for scientific communication. However, we definitely must state that this was not an easy decision."[37] A tradition of multilingualism in German, French, English, and Russian was now at an end. One year later, pursuant to this message, a second address to the membership appeared, in English, from the editor-in-chief.

> The long debates among the steering committee and council whether the [changes] will cause an "uprising" among the members or not were severe. Since the change one year ago, we definitely can conclude that this step was welcomed by the majority, even from the eastern [European] countries. Of course, there have been critical

> votes against this change which we considered seriously. However . . . we must take into account the international developments within the geoscientific community, so as to increase circulation of our *Geologische Rundschau* on the international level.[38]

Within three years, by 1997, the journal had changed its name and was publishing all research in English.

What about authorship? In *Geologische Rundschau*'s first two decades, nearly all authors were German, a fact not too surprising given the repercussions of World War I (German scientists were excluded from international meetings for years, while German scientific societies lost foreign members, and journals suffered from a drop in subscriptions and submissions). Starting in the 1930s, however, it was publishing authors from a number of countries—France, Austria, the Netherlands, Spain, Poland, Sweden, Bulgaria, Turkey, Romania, the United States, and China—all before 1950, with only a very brief lull in the late 1940s. Contributions in the 1960s and into the '70s arrived from many parts of the world, such as Latin America, South Asia, Indonesia, West Africa, and Israel as well. This has remained the case for the journal, at least to 2010. Anglophone nations, the United States and the United Kingdom in particular, have been fairly minor constituents, comprising about 5% of the total authorship.

What of collaboration? Figure 3.6, which plots multinational authorship for individual papers, reveals that the *IJES* has become particularly international in this sense as of 1990. It is a trend that, in part, directly reflects geopolitical events: after the fall of the Soviet Union, collaboration between European and Eastern Bloc geoscientists grew considerably. Those from Russia, the Czech Republic, and Poland are particularly well represented in papers of multinational authorship. In the early 2000s, meanwhile, papers with East Asian authors, above all those from China, become common. Indeed, papers both by Chinese geoscientists alone and in collaboration with European researchers increase markedly. By 2010, articles with three or more authors from different countries come to span a total of twenty-three nations, with single papers having authors from up to five countries. These numbers are double what they were at the beginning of the decade. There can be no doubt that with its transformation into an English-only periodical, the *IJES* has become more international than at any time in its hundred-year history.

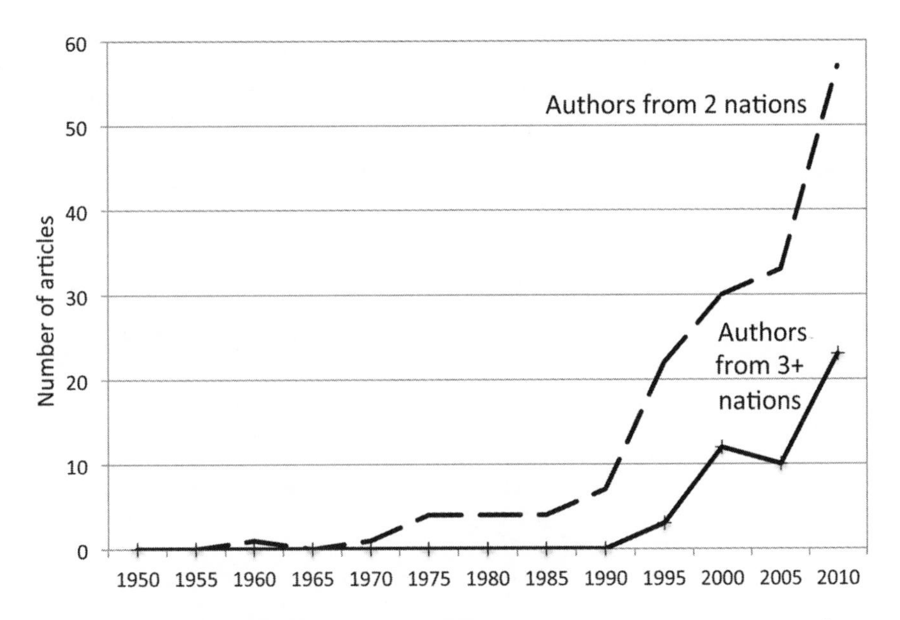

FIGURE 3.6 Level of international collaboration indicated by multinational authorship of papers in *Geologische Rundschau* (*International Journal of Earth Sciences* from 1997). Graph shows a rapid rise in such authorship after 1990.

The correlation is made clear by figure 3.7, which adds to figure 3.6 a plot of papers in English by scientists from non-anglophone countries. The overall correspondence between the use of English and the internationalism of authorship is striking. It is a finding that matches an important conclusion in the previous chapter, where we looked at the effects of past lingua francas in science: an international tongue is a pathway for broader, more diverse participation. It provides a medium for more scientists from a wider spectrum of backgrounds to contribute to the advance of any field.

No journal, to be sure, can be called representative of all science. But neither is the transformation of *Geologische Rundschau* into *International Journal of Earth Sciences* unique. To varying degrees, it has been repeated thousands of times. It is a history that presents not a microcosm but a sample, not an emblem but a specimen. Decisions in this history regarding language choice were neither imposed from above nor forced from below. They were guided from without, by the greater community of researchers, and generated from within, by editors—key gatekeepers—as

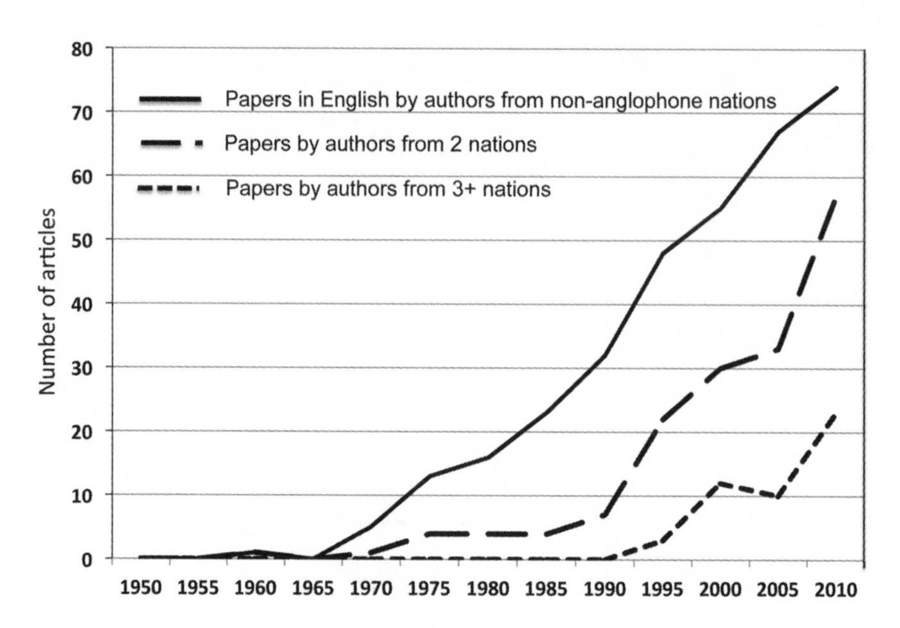

FIGURE 3.7 Graph showing trends in papers with multinational authorship and those written by non-anglophone scientists in *Geologische Rundschau* (*International Journal of Earth Sciences* from 1997).

they struggled and argued to bring their pages to as wide an international audience as possible.

Discourse Flexibility: A New Phenomenon in Scientific Publications

There is one final aspect to our look at English in contemporary science that deserves mention. We have examined number, distribution, and language in scientific publications (a subset of them, anyway), but not their actual *style*. After all, if there are now world Englishes to draw our attention, might there not also be different scientific Englishes? Or is the Anglo-American standard, as embodied for example in *Nature* and *Science* (the two journals cited more often than any others), obeyed as a kind of universal norm or desired goal across all fields?

The answer is complex, intriguing, and—to readers of this book—more than a little anticipated. Today, we do indeed find varieties of scientific English. Or to put it more accurately, we find acceptance of nonstandard forms of written English in a growing number of journals. Some of these forms are specific to regions where a world English has developed, and they share characteristics with that English variety.[39] But other nonstandard forms do not follow this pattern. They appear in the work of researchers from countries where English remains a foreign language. In these cases, divergence from the Anglo-American standard seems to reflect translation from a native tongue (for example, Chinese, Arabic, Italian) into English, or else an imperfect but nonetheless highly proficient command of English. A few examples, from psychology, epidemiology, and geology, provide some evidence.

> Copper (Cu) is an essential metal in cellular metabolism but also potentially highly toxic to fish. Despite of its essentiality, Cu requirements differ among species and even within different life stages of a single species. . . . Toxicity to Cu as well as other trace elements depends on species, age and diet, a reflection of variation in efficiency of absorption. . . . Heavy metals are generally accumulated in fish metabolic active organs in higher concentrations than in other tissues and organs.[40]

> Controlling infectious diseases has been an increasingly complex issue for every countries in recent years. Many scholars have investigated and studied lots of epidemic models of ordinary differential equations. The main aim of studying epidemic models is to help improve our understanding of the global dynamics of the spread of infectious diseases. . . . Moreover, considering some factors of the spread of infectious diseases, many authors in their literatures have investigated some new form of the models.[41]

> Seismicity of Egypt is attributed to the relative tectonic motion between African, Arabian, and Eurasian plates. Most of the seismic activity is distributed along three active margins: African-Eurasian plate margin, the Red Sea plate margin and the Levant-Dead Sea transform fault. Moreover, inland earthquake source zones in northern Egypt were reported by Abou Elenean (2007). . . . The identification of active fault planes in these seismogenic zones is essential for the potential seismic hazard that may carry on the closed urban cities.[42]

Native speakers from inner circle countries will find "errors" in each selection. Examples include: "Despite of its essentiality" in the first selection; "for every countries" and "authors in their literatures" in the second; "carry on the closed urban cities" in the last. In one or two cases ("for every countries"), grammatical rules are actually broken. But for the majority, this isn't true; it is more a matter of nonnative-speaker phrasing, deviations from Standard Written English. Readers will see that such phrases alternate with writing that obeys Anglo-American standards and would qualify as acceptable to any British or American editor. The major point, however, is that there are no serious problems of meaning. We even understand what the authors of the last sentence in the final selection are saying—that identifying active fault planes is essential for defining the earthquake (seismic) risk to Egyptian cities.

It would be facile to attribute the "deviations" in these passages to editorial sloth or incompetence. But the articles have all been peer reviewed and put through multiple levels of critique. Each journal has among its editors-in-chief and associate editors scientists from anglophone nations, including the United States. We can assume that the names of these scientists refer to real people who are not in a coma or catatonic state, and who therefore do their editorial jobs. It may well be said that whereas in earlier decades (1940s–70s) editors and reviewers took it upon themselves to revise articles so that they fully accorded with SWE, and that this is no longer possible given the volume of submissions, the impact over time is the same. What we see is a significant flexibility in what is now acceptable as written scientific English. And other factors are at work here besides an overload of journal submissions. As science has become more globalized, as more researchers from the outer and expanding circles described in chapter 2 have entered the scientific workforce, become authors and editors themselves, founded journals, taken seats on editorial boards, they have helped the global lingua franca to become less strict, more open to variation. The gatekeepers of many journals today seek to encourage a wider geography of contributors and to lessen language inequality between those who have English as a first or second tongue and those who do not, but whose work is valuable and deserves a wide audience. The passages above represent Turkish authors in an Italian journal; Chinese researchers in a journal published by an organization headquartered in Bulgaria; and finally scientists from Egypt, South Korea, and Saudi Arabia in a journal put out by one of the world's largest global publishing corporations (Elsevier).

What we do *not* see in any of these articles, or others like them, is a deviation from the standard form of argument in English scientific writing. Papers written about laboratory work follow the so-called IMRAD (Introduction, Methods, Results, And Discussion) structure, while those that involve fieldwork, theoretical investigations, mathematical modeling, methodological innovations, and mixtures of these obey organizational norms and styles of argument created during the twentieth century by western European science and refined in the postwar era by English-speaking researchers. These aspects of modern Western scientific discourse have been amply discussed in terms of basic rhetorical standards—the types of claims made by authors, the uses of evidence favoring these claims, the audience that is targeted, and the "suite of stylistic features" employed to give body and force to the structure of persuasion.[43] Such standards of English scientific prose—from the impersonal, nonliterary, and jargon-dependent style to the inclusion of elements such as the abstract, reference list, and normalized visual forms (maps, graphs, models, and so on)—act as universals, whether we are considering a medical study in India or a new paper in the *Brazilian Journal of Physics*. With minor variations, they are apparent in all English-language journals, including those published in nations where a particular world English exists.[44]

Flexibility in grammar and syntax is thus not matched by flexibility in rhetoric. At this stage in the evolution of scientific language, there certainly seems to be a true global standard in the fundamental forms of the research paper, forms that cross national, disciplinary, and even, in many cases, linguistic boundaries. What this suggests, rather strongly in fact, is that outer- and expanding-circle scientists have not made scientific language in English their own as yet; they have instead accepted with little or no amendment the rhetorical norms established by Western scientists, particularly native English speakers in Britain and the United States. Whether this will remain true or not can't be predicted at this stage. That speakers of various world Englishes haven't added their own innovations or made changes of some kind might lead us to expect this won't happen with speakers of English as a foreign language either. But such a conclusion seems premature.

Final Word about "Anglophone Science"

It is common today to speak of America's "decline" as the global leader in the sciences, and thus, by implication, English. Usually this "decline" is gauged by publications, patents, PhDs, or some other ready and routine measure. "The U.S. is no longer the Colossus of Science, dominating the research landscape in its production of scientific papers, that it was 30 years ago," says a recent report by Thomson Reuters (owner of Web of Science). "It now shares this realm, on an increasingly equal basis, with the EU27 and Asia-Pacific."[45]

China has four times the population of the United States, and in 2011 was at half its gross domestic product (approaching $7 trillion versus $14.5 trillion), having just surpassed Japan. It may lag America enormously in the scale, success, impact, and innovative vitality of its scientific culture. But this will not last. The Chinese government has declared it will "terminate" journals it considers weak, including thousands of Chinese-language periodicals in which plagiarism is rampant and in which little cutting-edge work appears.[46] China will succeed as a new global nexus of research, because it is adapting the Western model to its own needs, abilities, and circumstances (not to mention political culture). From investing large amounts of government money into building university programs to shifting more and more of its best journals to English, the nation is pursuing a known plan. At base, it has accepted the fundamental idea on which postwar American science itself was erected over sixty years ago. This idea was set out by Vannevar Bush (mentioned above), director of the Office of Scientific Research and Development, in his famous 1945 report to President Roosevelt, "Science—The Endless Frontier." Bush emphasized that research was not merely vital but indispensable to progress in the new era of world history. Only science, he said (meaning also engineering, medicine) could provide the tools and capability to "insure our health, prosperity, and security as a nation in the modern world."[47]

Bush saw science in America as a nationalist pursuit first, and a global one second. His view of the world was conditioned by the monstrous conflagration that had just ended, that had reduced most of advanced society to a smolder, and that left the United States looming over Europe and Japan like a paternal colossus. The notion of "American science"—or, a few decades later, "anglophone science"—seemed powerful and legitimate. Yet it would prove provincial and temporary in the long run. As researchers from more nations employ the English language to communicate their work and collegiality, science itself will become increasingly globalized beyond the borders of any single group of states, no matter how powerful.

Globalization in science therefore has another message. Despite all its tremendous achievements, modern technical effort has been at something of a frontier stage until now. This may sound absurd to those in wealthy nations. Here, the tidings of research and discovery are a daily item in the news, from the unraveling of the human genome to the discovery of new planets orbiting distant stars; Einstein and the Curies, meanwhile, lived a full century ago and have long been archived as icons. Modern science seems to be a mature intellectual industry, moving ever-more rapidly.

But to many hundreds of millions in the developing world, none of this has the ring of familiarity. In their countries, modern science as a domain of participation is really in its beginning phases. Until the end of the twentieth century, the most advanced portion of scientific work was confined to a fraction of the global community—roughly twenty or so rich nations, no more. The situation has now begun to expand in earnest. It is a change that scientists everywhere, and the world generally, should welcome without hesitation. More science, performed at a high level by researchers in an ever-greater number of countries, with a greater variety of specific goals, problems, and local knowledge than was ever possible in the past, can only be a good thing overall. In an interview regarding his Nobel Prize work on graphene, Andre Geim noted that there had been debates about what nationality or ethnic group should also claim the prize, whether British, Dutch, Russian, German, or Jewish. "To me," he said, "these discussions seem silly . . . I consider myself European and do not believe that any further taxonomy is necessary, especially in such a fluid world as the world of science."[48]

The twenty-first century will be the first to see something approaching a true world adventure in science, controlled by no one, belonging to all. In this new era, the value of a global language will only grow. Yet it's also true that this value comes with certain costs. These must be taken up, weighed, and evaluated. It is to this task that we now proceed.

(4)

Impacts

A Discussion of Limitations and Issues for a Global Language

Men have no right to put the well-being of the present generation wholly out of the question. Perhaps the only moral trust with any certainty in our hands is the care of our own time.

EDMUND BURKE

During the summer of 2010, I spent three weeks in northwestern Australia as part of a geology seminar run by the University of Washington. Our travels took us through a sizable portion of the Kimberleys, a wild, remote, and pellucid region of low-ridge mountains, sandstone gorges, and small towns encased by vast savannah ranches and bordered by tiny Aboriginal communities. In August, near the end of "the Dry," the area overflows with pitiless sun. The heat falls with a physical weight that can leave the visitor from higher latitudes breathless.

Native languages in Northern Australia are being lost at a disturbing rate. Long-standing conflict with white settlers destroyed many Aboriginal groups, forcing others into enclaves caught between a hunter-gatherer past now largely gone and a modern present yet to be fulfilled. One evening, during a rare, merciful breeze in the town of Timber Creek, I fell into conversation with Roger, an Aboriginal man perhaps in his late thirties, who had flashed me a friendly smile outside the grocery store and asked me where I was from. "I knew you wasn't from around here," he said. It emerged that he is a father of two boys, just as I am, and so we first traded tales of sibling relations and discipline problems. Then I mentioned the language issue; he nodded.

His boys, he said, knew three languages, none of them completely. It worried him. Roger's own birth tongue was Warimajarri, and he also knew Gurindji pretty well; but his wife came from over Tennant Creek way and spoke Warumungu and also Warlpiri. Both parents knew Kriol, a pidgin blending English and several Aboriginal tongues, used by white settlers and natives since the early twentieth century as a lingua franca across much of the region. The boys spoke their father's language fairly well, their mother's first tongue, Warumungu, better still, and Kriol best of all, because both parents and the extended family often used it at home. When the boys spoke Warumungu they might mix in some English words, too, since this was seen as status speech. At school, Kriol was used by the other children, and some of the teachers knew it. Roger noted that his sons and their friends typically switched back and forth between all three languages, depending on who was talking and whether or not a nearby adult might understand what they were saying. A big problem, he said, was that Kriol contains a lot of English words. "The whitefellas always thought me boys could speak English fine, but they can't. They had terrible time in school. Now the government change everything in school to English, so I think they learn it better. I think this might be good for them. Good for their future."

I asked Roger how he had learned his English. "I learned well at school—me mum made me study hard!" he laughed. "Later, I worked at a cattle station. Had to use it there. Now I been working in a store about ten year, using English every day." It was time for me to get back to camp, so I shook Roger's hand, wished his family well, and began to leave. "Hang on," he said suddenly. "Your boys in school, right? What they want to be?" "Well," I began, "one wants to be a doctor, the other a scientist, I think, to study the oceans." "You a lucky man," Roger shook his head and smiled, looking down. Then he looked up: "How many tongues your boys speak?" I replied, "Well, just one for now, though they're studying another in school." He gave me a humorous but pitying look. "Got to do better, mate! One never enough!"

Perspective

It is common to extol the benefits of using a global tongue for science. Such benefits are perceived to include not merely those of a short-term, practical nature, such as expanded collegiality and more immediate dissemination of findings, but those that will profit science in the long term through the globalization of knowledge. There is, too, the symbolic capital of a "global scientific community" (Rozier's "Republic of Letters" discussed in the last chapter) embodied in a shared global tongue. Most researchers around the world, therefore, if asked to comment, would be unlikely to find the status of English problematic or controversial.*

Yet international tongues, past and present, have not been universally kind to their foreign users. Such a language inevitably requires adoption, adaptation, and accommodation, none of which happen overnight, all of which involve difficulty and inequity. Former lingua francas of science—Greek, Latin, Arabic, Chinese among them, all of which we will discuss in chapter 5—did not attain their authority by consensus, but arrived on the back of conquest and empire-building. The impacts they had, for example on tongues that existed before their arrival, were often mixed. They could bring extinction to native languages in conquered territories, but also the creation of new tongues over time, such as the dozens of Romance languages that eventually emerged from Latin. Historically, a dominant language has had profound impacts on any preexisting intellectual community. It has altered many institutions of scholarly practice—education, literacy, practices of reading and writing, the definition of acceptable scholarship, the mobility of scholars themselves.[1] Overall, this has involved loss as well as gain. There are thus limits and drawbacks to be considered when a powerful lingua franca gains authority.

To what degree might this be the case with the use of English in the natural sciences? Are there important disadvantages that stand out, and if so, how serious are they? Given the many years of training and the intense competition for resources and rewards in contemporary science, can any such problems be addressed in some way? In truth, these are important,

* My own discussions with many scientists over the past decade, at international meetings and other encounters, show this to be the case.

even central questions for the future of scientific endeavor, since they involve the capabilities of researchers themselves.

Scholars of language have paid some attention to these realities and questions.[2] The role of English specifically and the issues it presents have been taken up or touched upon by specialists in the field of applied linguistics, yielding much that is valuable. Some of these authors are highly critical of English dominance in the sciences (seeing it, for example, as a loss of diversity); others are more drawn to its linguistic and social aspects. Be that as it may, a number of important conclusions emerge from this work. Here are the main ones relevant to our discussion:

- The global role of English in science has nothing to do with inherent qualities in the language. There is no wondrous "fit" between English and things scientific. The rise of this language is due to historical developments.
- English dominance is especially strong in the physical and life sciences and biomedicine. It is less pervasive in applied sciences and less (but increasing) in the social sciences and humanities.
- Such dominance is expanding because scientists want broad recognition, desire more opportunities, and understand that these now depend on publishing papers and citations in international journals, where English dominates.
- Since English is the native tongue for certain countries, researchers, universities, and companies from those places gain immediate advantage. Most anglophone researchers are monolingual and cite only papers in English.
- Poorer, developing nations have less capability to teach and learn English, so their scientists are at a strong disadvantage. This inequity can be understood as a situation of "haves" and "have-nots," or a type of core-periphery division, with wealthy nations at the center and developing nations at the margins.
- In many countries, most science is published in the native tongue. This work is internationally unknown, as it is not cited outside the domestic language community. Its scientific value may be local or it may not; we can't tell. Important work may go unnoticed, to the disadvantage of science and scientists everywhere.

- Many nonnative speakers of English suffer from low confidence when they use this language. There is much struggle, failure, and inefficiency attached to such use.
- Bibliographic databases such as SCI and Scopus, as a key part of the evolving "memory" of science, are biased toward publications written in English, though this has changed to a certain degree. Global visibility for any journal accepted into these databases is hugely increased; as a result, there is a powerful incentive for them to publish in English, creating a type of "feedback loop" for the further spread of this language in scientific publication.
- No imperial policy is at work to spread scientific English. Non-anglophone scientists are urged to use this language by historical realities, the policies of their research departments, universities, and companies, and the influence of their ambitions.

Each of these points, and others as well, will be taken up in what follows. The question of this book can't be adequately answered unless some of these conclusions are interrogated themselves.

The Issue of Equity

A global tongue seems to cleave the world in two—those who *can* use it and who *can't*. Or, more precisely: those who can use it by native advantage; those who can from years of disciplined or privileged study; those who feel they must forever struggle in a gray zone of inferior ability; and those who have little or no training in the tongue of power and are thus left weak and outside the wall of international scientific work, whether by choice or by lack of chance. In brief, though a world language provides new opportunity for scientists in a number of ways, it brings challenges and barriers for others. Even so, the simple "core-periphery" division is far too simple, as there are many levels of ability, partial ability, and inability.

Pointing out the problems of inability is not new. Even such a pro-English spokesperson as Eugene Garfield, founder of the Institute for Scientific Information (ISI), noted them some time ago.

> The fact that English is the internationally accepted language of research communication raises the issue of a language barrier in

> two senses. First, those whose native language is not English risk being unaware of—and overlooked by—mainstream international research unless they learn to read, write, and publish in English. Second, native English-speaking researchers risk being ignorant of significant findings reported in foreign languages . . . unless they become proficient in at least one other language.[3]

The world according to Garfield is a stark portrait: non-anglophone researchers *must* learn English, unless they wish to remain parochial; anglophones *probably should* learn another tongue.† Inequity seems stained by these lines. Nowhere does a concern over fairness appear.

The concern over linguistic injustice has been the primary issue raised against dominance of English in science. It is an issue that cannot be denied or ignored. Beyond any doubt, mother-tongue speakers derive some degree of advantage from the rapid globalization of English in science. Speakers with English as a foreign or "additional" language (EFL), or a fair portion of them, begin with a handicap. This is because the specific benefits for mother-tongue speakers are far from trivial. Besides the prestige that comes with fluency in the world's chosen language, these speakers, as trained researchers, are entirely at home in the required discourse and its vocabulary. Because they face the fewest barriers in producing it, they can more easily find entrance to its benefits (though this is never guaranteed), which include the rewards system in science. That so many of the world's international journals and scientific organizations, research institutes and conferences, university programs and scientific teaching, have moved to English-only status since the 1990s makes the point unarguable. Domestic journals in the United States, United Kingdom, Canada, or Australia become international by default. Native-like competence can, in some cases, aid the likelihood of getting a paper published and having it cited by others, thus improving one's standing and professional prospects.

But it's important to realize that the same advantages hold for any and all fluent speakers of English, not merely researchers from Britain or America ("native speaker," as we have seen, is a troubled category in view of those

† Garfield suggested Japanese or Russian, which were logical choices in 1990. Two decades later, different choices would make sense. There is an important point lurking here that will be taken up in the next chapter.

who speak English well as a second or third language). Endemic advantage goes to those with high competence in the language, no matter who they are or what their native language(s) might be. Another point to make is that mother-tongue speakers are not always quality writers or speakers in their own language, as experienced editors will gladly inform you.

Scientific discourse in any tongue is a highly formalized and specialized type of language use. One is not born with it and does not acquire it automatically or with ease, but must develop it as a skill over time, often with difficulty, as a distinct part of training. Among mother-tongue speakers of English, variability in this skill is immense and has long been so. There are therefore many anglophones who face big challenges in composing an acceptable scientific paper for publication.

But for EFL speakers with even less confidence and skill, the situation can be considerably worse. Writing and presenting can require a great deal of extra work and also money, if texts are sent out for editing and correction. Some journals will undertake such editing themselves, if they perceive the article to be of worth, but a majority lack this ability and so may well recommend that it receive outside help. Effort and resources are therefore drained away from other tasks, such as research itself (though communicating is essential scientific work, too). When added to the economic challenges commonly faced by scientists in poorer countries, these demands can seem discriminatory.

There are subtle hurdles as well. Linguistic conventions in scientific writing actually differ significantly among languages; verb tenses, use of adjectives and qualifiers, organization of articles, repetition of points, are all examples. If science written in French commonly employs the present tense (rather than the past tense, as in English), Japanese technical discourse is highly prolix (by English standards), while Chinese papers often include citations to renowned scientists whether they are immediately relevant to the work at hand or not. There is, too, the sticky problem of plagiarism in the published science of East Asia, where the concept seems to be either incompletely understood or poorly enforced. All these things add to the risk of error, misunderstanding, and even transgression in texts produced by nonnative speakers, even when edited by mother-tongue representatives. On the other side, some anglophone readers may view an article containing nonstandard ("odd" or "improper") vocabulary or awkward ("incorrect") phrasing as less worthy. In conference talks, poor or imperfect English, heavily accented or mispronounced, can frustrate listeners (who expect only first-language fluency) and lead some of them to associate imperfect performance with low importance.

More than fairness is involved, too. There are social effects to consider. Researchers who are more competent in English, who have studied and worked in anglophone nations, can find better job opportunities, to the point of gaining positions of authority in a nation's research establishment. Competence in English, perhaps even more than competence in research itself, may thus become a criterion for positions of power in a broad spectrum of scientific institutions, enhancing the status of the language still further. English skill is interpreted as the ability to connect with, and utilize, international science and all it can offer. Such a policy is logical in one sense, but will appear to be discriminatory as well. Beyond this lies the global need for language teaching and translation; nations with mother-tongue and fluent speakers of English gain multibillion-dollar industries in these two areas and so become, through their teachers and translators, a direct influence over the flow of language and the movement of knowledge within it.

Fears of loss are legitimate, especially in two particular areas. English is no serial killer of the world's tongues, as we have seen, but it is very definitely the one language in which new scientific terms are being coined, almost daily. Speakers of other tongues have the choice of adopting these terms from English, searching for an equivalent in their own language (yielding haphazard results), or not bothering. Over time, any of these options may weaken the ability of the target language to act as a medium for scientific knowledge, particularly in frontier research domains. This leads to the other area of potential "loss." Researchers who publish important work in their own (non-English) tongue risk invisibility at the international level. Their findings are never absent or mislaid; it is rather that the larger scientific community cannot see them and so remains ignorant of them. The drawback is twofold. Such proved to be the case, for example, with the avian influenza epidemic in Southeast Asia: important papers on the threatening nature of the disease "went unnoticed because they were published in Chinese-language journals,"[4] which also meant that the scientists who did this work weren't properly recognized for their contribution.

Issues of fairness therefore emerge from the real world. They are not all ivory-tower suppositions. They can be disputed in terms of their true extent and ultimate effects, but not in terms of their relevance and actuality. They do, however, raise questions of their own, and these shouldn't be avoided or dismissed either. Here are the major ones:

- How common, in actual terms, is bias against nonnative speakers in scientific publications? Given today's heightened awareness of global inequities and the need for change, is it routine that reviewers weight imperfect English more heavily than other criteria? Some statements by editorial board members suggest this may not be the case.[5]
- Where an article by an EFL speaker is returned because of "too many grammatical errors," how can we decide whether bias is at work or the manuscript isn't intelligible enough? Claiming that rejections for "poor writing" are, by nature, biased in the case of EFL speakers is surely wrong, as work by mother-tongue speakers is also returned for this reason.
- If the advantages attached to English now extend to any individual who uses it as a first or second language, or anywhere it has been successfully taught in school so that many adults are proficient (the Netherlands, Scandinavia), what does this imply about reducing the fairness problem elsewhere?
- The history of science shows that in every era in which a lingua franca held sway, intellectual culture did adapt, though time was needed for this to happen. Could the fairness problem represent a temporary phenomenon, something that—as David Graddol suggests, when he says that English will simply become a basic skill—will be gone by, say, 2040?
- Because any global language for science would bring the same issues of "injustice," what is the preferred option? Would it be more desirable for scientists to have several international tongues to choose from—say, Chinese, Spanish, English? Wouldn't a majority still need to learn a nonnative language?

The fairness issue, therefore, isn't at all simple. It raises difficulties of its own. It cannot be factored down to a doctrine of victims versus victimizers or a conspiracy among anglophone editors or a matter of "strengthening US global hegemony." Such characterizations are shallow, unhelpful, and usually wrong. Thoughtful critics understand and accept the status of English in science and the advantages it brings, yet they also look for ways to address the inequalities it creates. They have suggested an array of measures, not to mention solutions. By far the most common of these is to have journals provide financial or translation support for EFL authors.[6]

Journals, as well as academic departments and other institutions that can afford to help EFL speakers with their manuscripts (and with advancing their English skills), would be doing a good service for themselves and for science. This is clear. It defines an investment in human capital, one with inevitable returns. Yet we should not be sanguine about what is involved. Improving "troubled" manuscripts is labor intensive, akin to translation. Many periodicals and institutions do not have the funds or the personnel to properly support it or outsource it. At the most fundamental level, it returns us to the question of who should ultimately control the content of a manuscript. Leaving this responsibility entirely, or nearly so, with the author(s) has been the standard model in the modern era, despite the widely recognized truth that almost every paper emerges from a kind of in-depth collaboration involving grant writers, lab technicians, librarians, and others who comprise the human infrastructure of scientific work. Employing in-house editors and translators in this era of scientific globalization would appear to be a needed, adaptive expansion of this support system. But it casts into question the very idea of the "author" and who ultimately deserves credit for the published result.

Confined to the Margins

Related to fairness is a second issue of concern: the power of a global language to marginalize other tongues and their speakers. This can be as much a personal feeling on the part of individuals as an institutional, even national, matter of capability or policy. But like fairness, it leaves us with a residue of questions. First, however, let us address the concern.

A lack of confidence or a sense of insufficiency can be a dam to productivity for researchers engaged in high-pressure, competitive work. Scientists who feel insecure about their English ability can be less willing to share their work with the international community. They may be excellent researchers yet reluctant to write articles, give lectures, or speak at conferences and other gatherings where new findings are first revealed and (very important) vetted by verbal exchange among researchers. If forced to participate (by the rules of their institution or the demands of their career), they may spend many long hours on a paper or talk, taking valuable time away from the lab or field (or both). They may seek collaborators whose English is better than theirs, even if the collaborators were not part of the relevant research, thus diluting their own apparent contribution in any publications. They may even engage a ghostwriter or an editor who will help a good deal but leave them no more capable and no less insecure than before. Again, since research faculty are often evaluated by quantity of publications and their "impact" (level of citation by other researchers), any disincentive for submitting papers to international journals and symposia will have its own detrimental effects. All these factors on a personal level—as both survey and anecdotal evidence suggests—can give scientists the sense that they are operating more on the margins of their field, outside the mainstream.[7]

If such factors depend somewhat on individual personality, we also must reflect on the larger context. National wealth and dedication to R&D have a big role to play. Despite the building of new scientific capacity in many parts of the world, for example, frontier work in a number of fields remains the purview of a few rich nations. High-energy physics, space research, and areas of medicine and engineering are among these due to the Big Science factor—the scale and expense of related equipment and the infrastructure needed to sustain it. Cutting-edge science isn't cheap. Not every nation can (yet) afford it. Yet this does not at all mean that we can simply slice global science into the haves and have-nots.

The global scientific community can now be roughly divided into three domains: (1) countries wealthy enough to support all fields of science and engineering, or a large selection of them; (2) nations whose research capability is growing fast, but who must still select which fields to fully encourage; and (3) poorer nations able to support only a few fields, which tend to be applied science and related to that country's development goals (agriculture, water, energy, public health are mainstays). These groups are not bounded but grade into one another and are dynamic, evolving. The first group, no longer confined to Western nations plus Japan, has come to include countries such as China, Brazil, India, South Korea, Russia, and South Africa. The second group, by far the largest and most diverse, is led by states such as Mexico, Turkey, Taiwan, Estonia, and Poland, some of whom over the next few decades will advance into the first group or arrive at its edges. Finally, the third group comprises many of the least-developing countries, where even basic services such as electricity and clean water are not yet widely or reliably available.

Nations in the second and third groups work with limited resources, in other words. But not limited resources alone. More than a few of these countries, even if they show surging levels of gross domestic product, suffer a degree of backwardness from historical factors. In Vietnam, for example, the twentieth century saw a succession of Confucian, French colonial, Japanese, and Soviet controls; languages needing to be mastered by scholars of any stripe shifted from Chinese to French, then later to Russian. Scientists in these countries can find themselves on the periphery of contemporary scientific publishing almost by decree, faced with inadequate funding, inferior technical training, overburdened facilities, and even suspicions toward large parts of the scientifically advanced world, whose representatives regularly write and publish papers about these countries. Indeed, with regard to just one field—health-related science—an enormous publication gap has developed between researchers from wealthier states and those from poorer countries writing about their own homelands.[8]

Demands for the use of English can therefore add yet another layer to this structural context of marginalization. Such is confirmed by discussions with authors in developing countries, who list this as being among the problems they face for publishing in international journals. Editors of such journals, for their part, say that manuscripts from these authors can require more analytical work and editorial amendment than they are able to provide. The concept of a global program that would offer editorial help to authors in poorer nations (group 3, above) therefore seems to be something of considerable worth.[9]

Yet it would not solve the structural problem. Serious limits to scientific capacity, to computer and broadband technology, for example, will place any research program well away from the core of cutting-edge work—work that would be welcomed by international journals. Even if it were put into English, the research of scientists who suffer from such limits would likely be considered less publishable in competition with that performed at better-equipped facilities in more advanced countries. English is involved here as well. Language programs in group 3 countries, being fairly new or still rudimentary, often do not have enough qualified teachers and are less effective. Young researchers, not well taught in English, have a definite handicap in applying for both national and international grants even when the money is specifically for their ranks. Taking Vietnam as an example again, these problems aren't helped by the truth that young scientists who long for high-level training abroad (and who might serve as scientific leaders upon their return) are often ineligible because of poor English skills.[10]

The demand to use English can intimidate many authors in group 1 nations, too. Insecurity is widespread. It is not hard to imagine the following scenario repeated in a number of countries, particularly in Asia:

> The nervous [Japanese, Korean, Chinese, Vietnamese, Thai] postdoc spent two weeks creating slides, 30 hours drafting a script and 44 hours rehearsing. Altogether, she spent one month away from the bench so that she would not disappoint her supervisors and colleagues during a short informal presentation, in English, before her co-workers. Yet they remembered only the mistakes, she says.[11]

Our compassion is surely brought to bear. We readily conceive of an entire universe of small, daily tragedies like this. Yet two elements suggest we might be more discriminating with our sympathy. That the presentation is "informal" and for coworkers tells us that English is an integral part of the scientific culture here (every postdoc must give presentations as training for upcoming conferences). Moreover, mistakes are to be detected and discussed. Some members of the research team are English proficient, and demanding (science is not a profession for the weak of heart). Continued improvement is expected. Doing research at this institution means working in English as a baseline condition, understood by all. It is quite possible that our postdoc, if she survives the rigors of this particular setting, will be much more competent in the future (or intimidated into a different career choice).

Might we then, in this scene, be looking at a single, larval stage in a longer evolution? Anecdotes about struggles and "failures" are not the only stories to be told. How did the English-proficient ones get that way?

How long did it take them? Where and how did they study the language? Moreover, we shouldn't overlook the small fact that this "tragedy" takes place in one of the richest, most scientifically vibrant nations on earth. Saying that this proves how difficult the English situation must be misses the point. In a wealthy nation with a very high level of education and scientific instruction, becoming competent in a second language cannot be counted as an overwhelming barrier.‡

A greater tragedy befalls those scientists who are fairly or fully capable in English but marginalized due to economic and political factors—what we noted above. Japanese or German or Brazilian biologists, though dealing with insecurity in English, have the advantage of excellent training and a good salary, top facilities and a diverse university system, private-sector R&D of the highest order, and many international conferences held each year in their country. They consequently have a far better chance of a successful international career than their counterparts in Cambodia, Moldova, or Bolivia, where none of this is true. Indeed, a global language may well provide one of the *only* pathways for researchers in these poorer countries to collaborate, gain notice, and participate at a higher level in the global scientific community.

Another primary reason for this is that their native tongue may well be losing its scientific currency. As noted, a global language by definition pushes other tongues out of the international spotlight—think again of German and French, which together accounted for a third of all tracked publications in the late 1950s, but which fell to under 3% by 1996. This situation forces non-anglophone scientists to make choices about how, or whether, to incorporate into their own native tongue new terminology and concepts coined in English. Languages in which this is done successfully could remain scientifically vital indefinitely on a national level (both German and French are good examples again). Those tongues for which this doesn't happen will inevitably move to the margins, eventually replaced

‡ Cultural-historical factors can be very important, of course. The Japanese are not known to be highly successful second-language learners, due to a medley of reasons too complex to discuss here. Nonetheless, Japan's contribution to global publication in English remains above that of the United Kingdom, so it seems a good bet that a fair number of Japanese scientists have found a way to be proficient.

by English or another international tongue. As noted in chapter 2, in some countries, universities and scientific organizations have decided to install English for all technical communication. The Netherlands has gone the furthest in this direction, perhaps, with strong moves in Scandinavian countries, Germany, and more recently, Japan and South Korea.[12] Yet it also happens because of English's role as lingua franca. In Europe, research faculty and staff are now so international in many cases, with scientists and technicians from several countries (not always European ones) working together daily, that English becomes the only shared tongue.

What might this mean for relations between science on the national level and the public? The answer is not hard to see. Scientific language begins already several steps removed from ordinary speech; it is specialized, highly formal, rhetorically complex, dependent on a great many obscure terms, and difficult to comprehend for anyone outside the immediate discipline (true even for other scientists). Putting it into a foreign tongue removes it still further. The barrier between scientific knowledge and public discourse grows thicker, taller. Inasmuch as most governments aren't densely populated by scientists, this can impact public policy. Given that regimes in the world today, democratic or not, have identified R&D as key to the progress of society, any high walls placed around an understanding of science and its development would appear to be less than helpful. We might think of effects on schooling (writing of textbooks), media reporting, the level of informed debate and lawmaking about issues such as evolution, genetic modification, and nuclear power.

Progressive removal of tongues from domestic science is a loss. It is not merely regrettable but a true diminishment. Whatever international tongues scientific knowledge may have enjoyed in the past, it has always been culturally and linguistically plural and deeply nurtured by this. Nearly all the world's great scientists, whether in seventeenth-century Europe or ninth-century Baghdad, have been multilingual. This has been the normal condition for most of the world throughout history, to be sure, but it has especially been true of scholars, thinkers, and scientists. Having direct access to potent and useful thought in other tongues has always been an irreplaceable source of nourishment.

What questions, then, are raised by these realities of marginalization? The major ones would seem to be these:

- Are complaints about personal lack of confidence in English really more about the quality or extent of language programs? If we look at countries that teach English to their children from an early age (eight or nine) through high school and into university, some have highly successful programs (Norway, the Netherlands, Germany), while others do not (France, Spain, Japan). Can we consider the former to be "leaders" or even "models"?
- Isn't it legitimate to expect MS and PhD students, who have acquired highly complex knowledge in challenging fields of science, to be able to achieve competence in a foreign language (English)? Or does more of the responsibility rest with language programs and quality of teachers?
- Would it make sense to reconceive scientific training to include English as a required subject, no less than mathematics? Or would this prove to be yet another inequity, taking time away from scientific subjects and giving further advantage to mother-tongue speakers?
- A number of developing nations have shown rapid growth in publishing English-language papers between the 1990s and the 2000s (growth that is greater than any rise in the number of researchers). Obviously, *some* scientists in these countries, where marginalization is supposed to be high, are competent enough to use English regularly. Could the problem of "English insecurity" be decreasing over time?
- As with the issue of fairness, might it be that the problems of marginalization represent a temporary stage in the historical fulfillment of a global language in science? Is it possible that over the next several decades, as nations become richer and English becomes a routine skill, science will itself become multipolar (Europe and North America sharing leadership with East Asia, South Asia, and Latin America), so that monolingual anglophones become more marginal in importance, due to restricted language competence?
- Which, in the long run, is likely to be more marginalizing to a national research culture—the struggle to adapt to a global language or the refusal to do so?

Bias or Status?

A global language grows, in part, by status momentum. The very perception of the importance of such a language magnetizes new speakers, and suggests the advantage of institutional policies in its favor. Attached to one or more powerful nations, with whom many forms of opportunity are associated, the effect will be still stronger. At a certain point, official bodies, educational systems, and other mediatory sites take up the language, spreading it further, integrating it more deeply, compelling that it be used by future generations. Self-reinforcing cycles may therefore come to exist. Institutions that deal in communication can play a particularly significant and unexpected role.

For science, one such institution has done just this. The bibliographic database Science Citation Index (SCI) was created by Eugene Garfield in 1960. It is now available in greatly expanded form as the Web of Science, and covers over 6,650 of "the world's leading scientific and technical journals across 150 disciplines."[13] Garfield wanted to create a scientific way to study science (a "science of science," as it was described earlier in the century) that would involve quantitative means to track and map emergent fields, trends of collaboration, types of influence, and more.[14] Despite controversy, "scientometrics" grew in the 1980s and '90s with digital technology and requests for evaluating R&D performance in an era of fiscal accountability and concern over "competitiveness." Besides publication trends, SCI ventured into areas of "scientific worth" by tracking citation patterns, in the belief that such a value-laden idea was rationally constrained by communal choice and preference. The hotly debated metric here, "impact factor" (IF), a measure of the average citations per paper,[§] was introduced by Garfield as early as the 1950s, yet only in the 1990s did it become a widely employed indicator for individual papers, journals, and universities, consulted in administrative and policy circles

[§] Originally defined as the number of citations to papers in a particular journal divided by the number of papers published in that journal, for a specific time period. Over time, however, a number of modifications have been made to the denominator, to account for such things as article type (research paper versus letter versus review article, and so on).

and obsessed over by editors. In an ever-growing number of modified forms, it has been applied to institutions, departments, individual researchers, and nations, with highly disputed results.[15] Yet, as a metric, it has prospered. "The ability to judge a nation's scientific standing is vital for [those] that must decide scientific priorities and funding," wrote an influential author in 2004, who stressed that bibliometrics were a way "to measure the quality of research on national scales" by providing "metrics for . . . achievement." The author, we might note, is no naïve postdoc or bean-counting staffer. It is David King, chief scientific advisor to the UK government, writing in the journal *Nature*.[16]

By the late 2000s, citation data from SCI and also a new competitor, Elsevier's Scopus, were benefiting from a powerful array of applications that extended well beyond what Garfield and his followers initially intended. They also extended far beyond the borders of North America. Japan, Korea, Taiwan, and especially China were making use of citation data in evaluating their own scientific cultures and, in China's case, through the Shanghai Jiao Tong University, in ranking the world's top academic institutions.[17] Editors, scientist-authors, librarians (who make subscription decisions), and hiring committees, as well as science policy officials, were all paying attention to IF data by this time around the globe. In a number of fields—medicine, above all—IF had become fully integrated into publishing culture, bringing its own apparent impacts, with journals even manipulating their content to obtain higher ratings.[18] Researchers began to find themselves ranked, assessed, and funded partly on the basis of this data (despite cautions provided by ISI), especially in Britain. The rating of universities on a global scale using citation data, meanwhile, had become a small industry by 2011, with significant influence.[19] Some five decades after it was born, SCI was no longer an observer and analyzer; it had become part of the social practice of science itself.

Early on, ISI was convinced that English had become the dominant language of international science and would only grow in this role with time. Such was not a mere matter of blunt and unapologetic bias. In 1971, the institute decided to perform an analysis of journal citation patterns. Using data from the last quarter of 1969, it created a list of the most frequently cited journals and found, to its surprise, that a mere 152 out of 12,000 journal titles listed by articles in its database accounted for 50% of all citations. A miniscule portion of international publication turned out to have a massive importance. Of the 152, moreover, only 30 were journals from non-anglophone countries (mainly the Soviet Union, Germany, France, Scandinavia, and Japan), with about half of these published in languages other than English. When a second list of the 152 was assembled (for comparison) based on IF, the number from non-anglophone nations shrank to 20.[20] It therefore seemed obvious to ISI personnel that an overwhelming majority of "high-impact" journals were published in English.

Critics maintain that from the start, ISI has been biased in favor of English.[21] That SCI since at least the 1980s has claimed to identify the "top journals" in every major field of science yet selects only those publications that are either entirely in English or that contain important parts (abstracts, references, keywords) in this tongue gives some credence to the claim. Even in the case of "regional journals"—those, for example, in Spanish that have strong distribution across Latin America—SCI has demanded that bibliographic elements be in English. When compared with other bibliographic databases specific to particular fields, it is clear that ISI is highly selective and indeed can underrepresent the real activity of scientific advancement. Its general refusal to regularly admit such outlets as conference proceedings and, above all, open-access archives and online-only publications has been notable, though perhaps temporary. To an important degree, SCI has been highly conservative in tying its view mainly to traditional hard-copy science (and its online derivatives). No surprise that the comprehensiveness of its coverage has been declining over time—or that it seems weakest in some of the fastest-growing fields that employ the Internet routinely, such as computer science.[22]

The total result is that researchers who write only in English therefore have a better chance of a large presence in the database. Likewise all-English journals; these, too, will be well represented. Yet in any nation with an international tongue or a deep scientific tradition, there is much incentive to write and publish in the native tongue as well, to be an integral part of a thriving domestic landscape of research. Authors in these nations therefore can suffer the fairness issue, as well as the need to find a dynamic balance between national and international reputation. This will not have been greatly aided by the growing influence of SCI itself. As its indicators have been put to greater use by those in positions of power over funding and hiring, it has had its own impact on the trend toward all-English publication. Small surprise, then, that anglophone nations appear to be consistently overrepresented in the results of various measures (for instance, papers per researcher, per university, per country).

Not all measures, however. When gauged in terms of articles per million inhabitants (or per capita) for 1998–2000, Switzerland was seen to lead the table, and the Scandinavian countries (Norway, Sweden, Finland, Denmark, Iceland) came out particularly dominant.[23] Meanwhile, IF scores calculated for 2005–8 also show Switzerland to be highest, followed by Denmark, with the United States third. Though the United Kingdom, Ireland, and Canada were in the top ten, so were Norway, Sweden, Belgium, and the Netherlands.[24]

Northern Europe plus Switzerland, therefore, emerge as the true global scientific leader, with the United States, the United Kingdom, and Canada less important overall on a per-researcher basis. This is interesting. It offers a portrait of world science very different from the standard oppositional triad posing North America versus European Union versus East Asia. These "new" leaders are the most multilingual nations in Europe—well-educated Swedes, Swiss, Danes, and the Dutch, for example, commonly know three, even four languages. One of these is English.

To a certain degree, Thomson Reuters has sought to address the question of language bias. Since 2005, it has expanded the number of non-English journals in its database and has aimed to begin coverage of national science as well (it has also begun to include books and conference proceedings). In the latter case, Thomson Reuters has formed a partnership with the Chinese Academy of Sciences (CAS) to create an entirely separate citation database for Chinese publications. In 2011, this new database included twelve hundred scholarly journals that the CAS helped to recommend, with article information in both Chinese and English and about 40% of all papers carrying English-language abstracts. As a potential model for other countries, this implies not only the preservation of national science (selective though it may be) but also the removal of its invisibility to the international community. It is a tempting thought, that the single institution presumably most guilty of globalizing English in science becomes a key preserver of the multilingual dimension to world science. Could it happen?

We are left, nonetheless, with several questions:

- Is it fair to say that databases such as Web of Science and Scopus, and the indicators based on them, really refer only to international science? International science may have the higher status, yet it isn't the whole picture by any means, as we've seen even in frontier investigations.
- On the other hand, how firm is the boundary between "international" and "national"? Could it be weakening, as it seems to be in some areas (northern Europe), where English is being substituted even for native tongues in science? Could SCI / Web of Science be contributing directly to this boundary dissolve, thus to the loss of national languages in national science?
- To what degree has SCI / Web of Science actually added to the spread of English in scientific work? Is it a primary or secondary (or tertiary) agent? If significant, is its influence widely distributed across the whole of science, or concentrated in certain fields and ignored in others (and why these)?
- If English has become the global tongue of science, to what degree does the purported bias in SCI / Web of Science matter any longer? Are there a significant number of researchers that reject this role of English (if so, bias would indeed be an issue)? It might be noted that researchers who have found fault with the use of IF and other "quality" indicators rarely, if ever, bemoan a pro-English slant in the database.[25]

Endangerment of Other Scientific Languages

English has not wiped out the world's tongues, but it is the one language whose growing use in domestic research around the world can reduce scientific communication in other languages. It is the tongue that can replace national languages in this context. It is also the language in which new standards are set in scientific discourse, especially where new vocabulary is coined (a central process in science), older terms questioned, and vestigial ones replaced. Moreover, English acts as the testing and proving ground for any new technical vocabulary created in other tongues—if a term is not adopted into English, its chances of survival are poor.

An international language in a particular sphere of communication thus works against linguistic diversity in that same sphere. This is known as domain loss in linguistic circles. It defines a phenomenon that has occurred innumerable times over the course of history. Indeed, all international tongues succeed by this process, to ultimately varying degrees. In Japan and Korea, for example, classical Chinese was chosen as the language of writing for centuries, until each country developed its own orthography, at which point Chinese retreated mainly to the domain of written scholarship. Arabic, on the other hand, served a kindred purpose in Persia only to a limited degree (mainly in religious matters) and for a limited time. Languages installed by means of conquest have commonly taken over the domains of government administration, military matters, law, tax collecting, and some areas of engineering (road building, for instance). This has not at all meant the inevitable death of previous dominant tongues, as they have continued to thrive and expand in other domains, not least vernacular ones. Yet it has meant a certain loss for such languages, which are no longer able to satisfy every linguistic need of a community.

There is no doubt that the global role of English in the sciences puts pressure on other tongues to stay current. Researchers, editors, and other communicators in national science must choose whether to adopt new vocabulary directly from English, search for an equivalent in their own language, or, finally, to use English itself. These choices, through the broader topic of domain loss, became the subject of heated debate in Scandinavian countries, above all in Norway, Sweden, and Denmark, during the early 1990s.[26] Behind the controversy lay a number of factors, such as worries over national identity in the face of the European Union and, more specifically, ongoing linguistic questions in the public context, such as the struggle over standardizing Norwegian (a relatively new official language, since the country achieved independence from Denmark, and Danish, only in the nineteenth century) and achieving a single, understandable form of spoken Swedish. Another factor, however, was a fear of language endangerment—the concern that if linguistic territory were ceded to English in so important an area as research and training, all domains would eventually follow, and the national tongue would become extinct.

By the late 2000s, the controversy had largely cooled. Nearly twenty years after worries were first expressed, Norwegian, Swedish, and Danish were not threatened, despite the ever-increasing use of English. Scientific work and publication in each national tongue hadn't entirely disappeared either. Rather, a certain accommodation had been reached. Preserving each of these languages as a form of national identity remained a theme in cultural policy (strengthened by concerns over immigration), but it lacked the edge of anxiety it had attracted earlier, in part due to comments in the media by linguists themselves.** Finland, which hadn't engaged with such anxiety very much (possibly due to its long-standing use of Swedish), had moved ahead with English implementation and a highly successful effort at internationalizing its companies and universities. Positive attitudes toward English are the rule here.[27] Sweden, dealing with immigration and a "new multilingual nation," passed a national language policy in 2005, declaring language rights to all groups and Swedish the national tongue. Yet at research universities in the country, both official and unofficial policy often put English above or on a par with Swedish.

With regard to scientific research itself, forms of accommodation vary. A good example is provided by the Geological Survey of Norway,[28] a government research agency under the Ministry of Trade and Industry, whose technical publications are mainly in Norwegian (approximately 65%) but change to English when they concern areas outside Norway and involve topics with international relevance or application. Geological papers and reports in Norwegian show evidence of having both adapted English terms and employed Norwegian equivalents, so that national geological discourse has remained fully current. Publications by university researchers in certain other fields, such as genetics and biomedicine, are exclusively in English at this point. The same is largely true for most fields in physics and mathematics. This isn't surprising, since the faculty in each of these disciplines is itself international, and many research projects are now undertaken in collaboration with teams from other countries. It would appear, therefore, from an informal survey that the degree of domain loss is field-specific and closely related to the immediate international dimension of research.

If we ask how scientific work in Scandinavia is represented to broader domestic and international audiences, a somewhat different linguistic picture emerges. A good example here is Denmark's Agency for Science, Technology and Innovation,[29] which oversees national research activity,

** A good example of this can be seen in the article "English No Threat to Danish," *Copenhagen Post,* April 20, 2011.

funding, and policy; the papers, reports, and other material it makes available are in English and Danish both if judged to be of potential international interest, but Danish only if aimed at informing Denmark citizens. The same type of approach can be found in Sweden and Norway, for example in the various publications of each country's National Research Council.

In short, domain loss in science and science-related communication has been significant but far from total. An evolving balance or semi-equilibrium seems to have developed in the use of each tongue vis-à-vis English. This leads to several questions about the long-term nature of the situation.

- If domain loss does become complete in some fields of science for various nations, will this mean that the same will occur in other fields over time? Would it have a negative, positive, or neutral effect on the research that is being done in those fields and nations?
- Is it more helpful to view the situation of domain loss as a local problem for each national tongue, or as an impact from science's own globalization?
- In any single country, but especially a multilingual one, would linguistic diversity in science lead to greater freedom—or fragmentation?
- How much of a guide might the history of science be here? How relevant is it that the development of science in a variety of cultures—ancient Egypt, Greece, India, China, Islamic Empire, medieval and Renaissance Europe—depended to a considerable degree on the existence of a lingua franca?
- Would linguistic diversity in science guarantee diversity of thought, perspective, methods, approaches—a wider range of creative mentalities? Or rather, does arguing on behalf of linguistic diversity require us to resolve, once and for all, the Sapir-Whorf hypothesis (language has a determining influence over thought)?

Summing Up: Risks from a Global Language

So what type of assessment seems justified? Are the issues—fairness, marginalization, pro-English bias, and domain loss for other tongues—real and worthy of concern?

The answer must be yes. The issues are real, and they affect many thousands of scientists and students today in dozens of nations. It is not hard to imagine, in fact, all the mentioned issues working together—the sense of injustice due to one's lack of English skill and limited publication of one's work and career possibilities, leading to a lack of confidence and marginal research performance. No one should dismiss, moreover, the more subtle consequences. For some time, as an example, researchers from poorer nations have tended to ally themselves with colleagues in anglophone countries most of all, especially the United States, hence decreasing their collaboration with other scientists from nearby developing countries that share similar research agendas and therefore aims. Another disturbing consequence that must be pointed out (but doesn't fit into the categories of this chapter) is the impact on libraries. During periods of stagnant and shrinking budgets, as have existed for Western libraries since the 1990s, librarians have been forced by institutional policy to cancel subscriptions for underused material, particularly that in a foreign language.[30] In anglophone countries, especially the United States, where few scientists refer any longer to non-English publications, this could well include valuable research publications in many tongues, even major ones such as Portuguese and Arabic.

Yet this disturbing image can't be counted as the alpha and omega of the global situation. Common as the aforementioned issues may well be, they are far from definitive or universal. Moreover, they may be decreasing in importance.

As our discussion suggests, they are also endemic to the situation of a global language. It wouldn't matter which tongue sat in this high seat, whether English, Chinese, or Yoruba; aspects of all four issues would be brought to bear. Nor would it really matter if there were several global tongues of science, since a great number of researchers would still be forced to learn at least one of them. During the past two centuries, before the rise of English, most of these issues undoubtedly were generated by other international languages, such as French, German, Spanish, and Russian. It is not hard to imagine, for example, young American or British chemists in 1885 sweating over an article they wished to submit for *Berichte der deutschen chemischen Gesellschaft* or, with wrinkled brow and remedial German, struggling through one of Emil Fischer's famous lectures on sugars at the University of Würtzburg. Many similar situations undoubtedly took place, though with a sharper political edge, in eastern Europe and the Baltic states, where Russian was the de facto language of scientific training for several generations. The difference today is that an immense diversity of nations, at many different income levels, are involved with scientific English. Which brings up an essential theme.

Linking the issues of fairness and marginalization in another way, in terms of cause, is a single factor—language teaching and learning. Of course, this "single factor" is anything but simple. Language instruction and its successes or failures are part of a nation's educational system and reflect the outlook, priority, and resources devoted to this system in general and the instruction of foreign languages in particular. Models of instruction, benchmarks of achievement, the training and support of teachers, and the actual functions of a specific language in the relevant society are all involved. Some countries, particularly the most multilingual in Europe, have made skill-building in English by science students and professional researchers highly successful. In most of these places (such as the Netherlands, Norway), English has a daily presence in professional society; those who study the language have a high probability of using it when they enter the job market. In a number of other nations, such as Mexico and Brazil, scholars and professionals are able to acquire high-level English, too, though more unevenly. Yet in countries where a single national tongue tends to dominate training and work as well as the educational system in general, such as Spain, Russia, and Japan, English teaching and learning tend to be less effective. Here there is less opportunity to use the language on a routine, daily basis. Then there are the many poor and less wealthy nations where a combination of historical hurdles and resource problems, including lack of qualified teachers and teacher training, can make it difficult for scientists to learn English well.

These are rough-and-tumble characterizations, to be sure. They carry enough of the truth, however, to show that while a global need exists for scientists to know English, the efforts to address this need via education have highly variable levels of success. The problem can't be solved easily or quickly; aside from the monetary and manpower aspects, there is the noted place of actual English usage in each particular society—if common, it can help strongly to encourage higher levels of proficiency (there's no substitute for actual use of the language outside the classroom), but if rare or absent, such encouragement can be low. There is also the truth that education tends to be a battleground of competing visions and interests, and decisions made under one government may not last (example: Malaysia's embrace of English in science and engineering during the early 2000s; then, under pressure, its decision to reverse this position and banish the tongue from such training; and then a reversal in more recent years back to its use). Moreover, there is cause for realism. Improvements in English-language teaching and learning are only part of a needed system: giving young researchers strong skills in the language will serve them best only when there are also research funding, laboratory facilities, and domestic industries that might employ scientists.

This being said, the need to know English for any level of international communication, and the doors that this ability can open, are profound, undeniable, and growing. Making language study, even as "scientific English," a core element in technical training would appear to be essential, despite the hurdles. It shouldn't be thought excessive to ask that students who reach a graduate level in a field as complex as geochemistry or molecular genetics be able to acquire functionality in a foreign tongue (a point that should be applied, no less, to monolingual anglophones).

Change is clearly happening. We see it in the rise of English-language publications coming out of developing countries around the globe—Argentina, Poland, Turkey, Thailand all providing strong examples (see chapter 5). Somehow, without the benefit of divine intervention, researchers in these places have found ways to be competent in English. Admittedly, this may involve many things, from the use of mother-tongue "editors" to actual ghostwriting. Yet much evidence, such as the many non-anglophone scientists at international meetings who present and discuss their work in English, suggests that the level of comfort with the language among scientists worldwide has risen greatly in the past decade.

This is what we should expect, in fact. The history of lingua francas in science does indeed recount, again and again, the tale of intellectual cultures adapting to new internationally dominant tongues—whether Arabic in Persia or Chinese in Japan. Such adaptation has never come without struggle and cost or without the requirement of time. But it has happened with an understanding (forced or otherwise) of the necessities and benefits involved.

Global scientific English defines a dynamic reality, and therefore an evolving one. Its ground-level disadvantages observed in the 1980s, 1990s, and early 2000s are no longer viewed as so pressing. Thousands, perhaps tens of thousands, continue to be their unfortunate subjects, true enough—and for these researchers, many of whom undoubtedly have valuable knowledge to share, forms of help should be made available, for the sake of fairness and of science. But it is also true that many thousands are moving beyond such a situation. Some earlier criticisms of English-only science aren't true any longer. It isn't the case that editorial boards of international journals are heavily populated by Americans and Brits, with a peppering of western Europeans, as they once were. It isn't true that nonstandard English in a research paper counts as a universal reason for rejecting the manuscript, nor that such papers are always edited into "perfect" Anglo-American style. It isn't acceptable anymore that talks and papers by researchers from developing nations be considered of lower importance. And it no longer happens that scientists from developing nations always look to partner with those from the United States or Europe; regional collaboration is on the rise. Finally, and no less meaningfully, it isn't at all a surprise any longer for scientists in the West to encounter colleagues from Asia, Africa, or Latin America whose English is proficient, excellent, or entirely fluent.

In the end, drawbacks and limitations must be weighed against benefits and opportunities. If the former are real (and they are), they require an audience beside the latter. Each brings a series of caveats before the jury. It is these that we will finally examine, with a dry and lidless eye, in the next chapter.

A Final Story, and an Idea

Toward the end of my stay in the Kimberleys, a remarkable event occurred. Driving back to the Gibbs River Road from Tunnel Creek (a kilometer-long walk through a magnificent limestone cavern), the left back wheel of our jeep suddenly flew off, bounding into the bush like a frightened wallaby. The vehicle ground to a stop in the red dirt. Pieces of broken metal lay about, like shrapnel in the dust.

Within a short time, several cars and jeeps had stopped, and we found ourselves surrounded by friendly voices. We were plied with water, beer, bananas, energy bars, and a flow of encouraging words. Two men proved to be mechanics, and conceived a repair that took the better part of two hours. The sun drove the rest of us into the miserly shade of a boab tree, where I ended up talking with a leathery-faced young woman wearing a broad sunhat who was a grad student in environmental science at the Australian National University. I remarked on the kindness shown, and was told that this was typical in the outback. People like a gathering, she said, and also stop because a breakdown can be fatal in such a remote area. The Aboriginals, she noted, had many ways to find water in the Dry. Their territory was like a secret text; they knew hidden watering holes, rainwater gouges in granite, the hollows of certain trees, hand-dug wells in hidden, shady places. In times of drought, they used plants and could find new water by observing the movements of birds. Early in the era of white settlement, men or boys were sometimes kidnapped and forced to show where water could be found. This was especially true of expeditions to the interior, but not those of Donald Thomson.

Thomson was an anthropologist, perhaps the most famous champion of Aboriginal causes in Australian history, who led a series of ethnographic trips between 1957 and 1965 to the Bindibu people in the Great Sandy Desert. Even at this date, many of the Bindibu had never seen a white face and lived the hunter-gatherer life as they had for tens of millennia. To the whites, the Great Sandy was "the land where men perish," the harshest desert on the continent. The first to cross it was Colonel Peter Warburton in 1873; his expedition fell into legendary difficulties of heat, thirst, and incompetence, and members were forced to survive by eating their much-abused camels and by relying on their Aboriginal tracker, Charley. Several subsequent crossings by later explorers all suffered greatly from lack of water. Thomson, despite many preparations, did not avoid this challenge. Though he had only a single companion plus two trackers, they promptly lost one of the jeeps, and lost the second one before long in the midst of a raging sandstorm.

Thus stranded, Thomson was soon befriended by a group of the Bindibu. They showed absolutely no fear of him or anything he did, and this extended to the children, who would come over and squat to watch whatever task he was engaged in. Learning a portion of their language, teaching them some of his, Thomson was allowed to live and hunt with this group for several months. He found the people not merely sociable but complex, witty, ingenious. He saw that their lives and songs revolved around water. "It soon became apparent," he says in his account, "[they] had a practical knowledge of the ecology of the desert in advance of any white man. . . . There was certainly a great deal more water . . . than had been assumed."[31]

Eventually, a replacement axle for the second jeep arrived. Thomson decided it was time to return. On the eve of his departure, the people gave him a gift. It was a type of life-giving knowledge. Carved on the back of a wooden atlatl was a design of lines connecting a set of spirals that the leader, Tjappanongo, showed to him.

> Sometimes with a stick, or with his finger, he would point to each well or rock hole in turn and recite its name, waiting for me to repeat it after him. Each time, the group of old men listened intently and grunted in approval "Eh!"? or repeated the name again and listened once more. This process continued with the name of each water until they were satisfied with my pronunciation. . . . I realized that here was the most important discovery of the expedition—that what Tjappanongo and the old men had shown to me was really a map . . . of the waters of the vast terrain over which the Bindibu hunted.[32]

Thomson's journey can be read as an example and as a parable. A more capable society will be one that is able to utilize directly the knowledge of others. In the future, knowing English alone will not be enough. As more of humanity learns this language, and therefore becomes still more multilingual (in many cases), monolingual speakers will increasingly find themselves approaching the edges of a desert where they have less access to a greater part of the outside world. The greatest long-term danger coming from the global spread of English—could it be to its own native speakers?

(5)

Past and Future

What Do Former Lingua Francas of Science Tell Us?

Not to know what has been transacted in former times is to be always a child. If no use is made of the labors of past ages, the world must remain always in the infancy of knowledge.

CICERO

Adelard was born in Bath, southwest England, probably in the year 1080, at a time when the country was in great disarray. Barely a decade before, William the Conqueror had invaded, casting out the nobility, installing a foreign aristocracy, and reforming the clergy. Speech and custom were everywhere in flux. A new language had been imposed on the upper levels of society, and new words and phrases were already beginning to pour into the native Anglo-Saxon tongue. Politics, the military, law, commerce, and even the church were now beyond reach without knowledge of French. Latin, the language of learning and liturgy, was a saving grace. All who had some learning could speak it with some proficiency, while teachers and the clergy who commanded the domains of scholarship used it as a second tongue.

Bath was where the Romans had turned Britain's only hot spring into a health resort. A millennium later it remained famous, visited by people from all over England. The town was therefore a center of stories and rumors and news, about Britain and the Continent. Adelard was only seven when troops fighting an insurrection against William's successor came and set the town on fire. The new king, William Rufus, recognizing the loyalty and importance of Bath, put it in the care of a renowned physician and bishop, John of Tours, who helped in the rebuilding and erected a great cathedral and a new school. Within a brief time, Adelard proved himself a brilliant student there. He was later sent to Tours, a center of learning in the most learned nation of the day, having mastered Latin, French, music, science, and mathematics while still a youth. He became a teacher himself, in the school at Laon, founded by the great Anselm. This, however, did not satisfy. Even while still in his hometown, he had been excited by stories from crusaders. Adelard left France in 1109, traveling first to Salerno, with its school of medicine, then to Sicily and Greece.

His travels then took him to North Africa and Asia Minor, where he sought "the wisdom of the Saracens." He spent years learning Arabic, studying the texts of Islamic science, and made the decision to translate these works into Latin, to give to Europe the great riches for which he had searched. Euclid's *Geometry* entered the West by Adelard's hand, as did works by al-Khwarizmi introducing Hindu-Arabic numerals. He wrote on the abacus and the astrolabe, and became known for his *Natural Questions*, a textbook about the physical and animal world. Today, nine hundred years later, he is called the first English scientist. Yet Adelard belonged not to England but to the early decades of the great twelfth-century renaissance, and to the transformation of Europe. It is always best, he said, "to visit learned men of different nations. . . . For what the schools of Gaul do not know, those beyond the Alps reveal."[1]

A History of Many Tongues

No attempt to understand the present and the future roles of English in science would be complete without a detailed look at the past. Previous lingua francas of science provide an important, even essential, body of evidence. Indeed, such languages constitute the *only* real evidence of this kind, and so demand examination and reflection. Otherwise, we are without context, on a planet lacking an atmosphere.

Periods in which knowledge about the natural world advanced to a high degree were characterized by major lingua francas.[2] Many today would think immediately of Latin in this way, for it was the tongue of educated Europeans well before the lifetime of Adelard and long afterward, from the sixth to the sixteenth century.[3] Yet, as Adelard's story tells us, Latin was but one of the great linguistic wheels by which science moved forward. Arabic was another; so were Greek, Persian, Sanskrit, and Chinese. The history of science is also a history of languages.

This historical pattern changed for a relatively brief time at the beginning of the modern period. A three-hundred-year era of modern scientific achievement, from about 1680 to 1970, could even be termed an anomaly—it stands as the only period of major advancement *without* a single, true lingua franca. By the late 1600s and 1700s, that is, a Europe fragmented into competing nation-states was hurrying the death of Latin as the regional tongue of knowledge, replacing it with the vernaculars, which now came to dominate—Priestley and Darwin writing in English; Lavoisier and Pasteur in French; Gauss and Einstein in German. There were temporary shared languages for science, to be sure—French filling the gap left by Latin on the European continent (not in Britain) for much of the eighteenth century, German serving a similar purpose for a more abbreviated time in the late nineteenth and early twentieth centuries. Latin even continued in a few fields, especially medicine, for parts of northern and eastern Europe. Yet these examples were wholly unlike the vast purview that tongues such as Arabic and Chinese commanded over their respective regional domains for many centuries.

During the decades between 1970 and 2000, this era of national-language dominance came to a close. Were Marie Curie and Albert Einstein to have sat at their writing desks in 2010, they would have been compelled to compose in English. Einstein, in particular, would have had no choice: his favored journal, *Annalen der Physik*, where he published his most epochal work, including the great papers of 1905, the so-called annus mirabilis, is now an English-only periodical (though retaining its original name). National science will not disappear, of course, as we have said—to those who fund and manage research, it remains vital, even aggressive, linked to such national priorities as prestige, defense, economic competitiveness, public health, and more. But for scientists themselves, nationalism isn't the factor that drives research. As noted and emphasized earlier, researchers everywhere embrace not only internationalism but globalism as an ever-growing dimension of their work and the future of their disciplines. It is a dimension apparent in a great many aspects of contemporary science, such as the multinational authorship of papers; the growth in international conferences; the recruiting of technical talent from many countries by R&D companies; the rising number of foreign students from around the world; the global mobility of academic researchers; and the daily use of the Internet for communication and for creating global databases. All of this, as we have seen, means use of English.

As an approach, looking for shapes of the future by studying those in the past makes sense, but comes with a caveat. Reassembling linguistic history can create the illusion that we are brushing the soil from bedrock universals: what happened to Greek or Arabic will inevitably befall English. But, of course, this isn't true—the past offers clues for informed conjecture, not certainties for expectations. Essentialism, as historians so often point out, is a lure to counterfeit finalities. No previous age has known the globalizing power of digital communication, for example. Using the past therefore begs a degree of speculative boldness tempered by the knowledge that we are doing humanity's work, not God's.

What follows is a look at four major lingua francas of science past, with a view to their impacts and what might be gleaned from them as possibilities for the future of English in science. My choices are limited to Greek, Latin, Arabic, and Chinese. Other tongues would obviously be worthy. But the four I have selected were especially widespread and contributed much to the world's scientific. What, then, do they tell us?

Science in Greek: Classical and Hellenistic Eras (Sixth Century BCE to Third Century CE)

Greek was the lingua franca of the Eastern Mediterranean region for no less than seven hundred years. It was certainly dominant during the classical and Hellenistic periods, from about 550 BCE to 200 CE, and in Asia Minor, site of the Byzantine Empire, much longer. Greek science began in the sixth century BCE, following an episode of colonial expansion and the absorption of other peoples.[4] This expansion brought within the orbit of Greek language and culture portions of Macedonia, Ionia (western Turkey), southern Italy, Sicily, North Africa, and most of the islands in the Aegean. The hotbed of new inquiry into nature, based on rational theorizing and observation, was Ionia, especially the port city of Miletus, which had commerce with other parts of the Near East and probably with North Africa as well. Based on the mathematics of Pythagoras (born in

Ionia, on the island of Samos, about 570 BCE) and on Greek astronomy, it is clear that intellectual influences had arrived from Babylonia and Egypt, as well as from other peoples of the interior region to the east. Such influences were probably crucial in the beginning of the era and were soon transmitted to other intellectual centers, such as Athens, via the Greek language.[5]

What then occurred, from the fourth century BCE onward, was by any account an astonishing period of advancement lasting for two full centuries, beginning with Hippocrates, Aristotle, and the young Euclid, and continuing down to the time of Hipparchus (second century BCE). Except for Aristotle, the larger part of this science was concerned with mathematics, mathematical astronomy, medicine, and alchemy (chemistry), and in Strato and Archimedes a number of areas in physics and engineering. Aristotle, however, adds to the curriculum a host of other subjects—biology, botany, zoology, meteorology, geology—though certainly not as individual fields. During this period, the Greek language spread throughout the Near East and Egypt by defeat of the Persians and the subsequent conquests of Alexander the Great. By the early third century BCE, Alexandria had become a thriving intellectual center and the site of the Great Library, which was rivaled by the collection at Pergamum in Asia Minor.[6]

Rome took control of Greece in the second century BCE. Even so, Latin never came to dominate the Eastern Mediterranean. In fact, Greek was maintained as the lingua franca even as an administrative tongue in the Eastern Roman Empire, and it was left entirely untouched in intellectual matters, particularly science and philosophy and in the schools. During the years of the Republic, Roman scholars and wealthy aristocrats learned Greek to gain access to what they considered a superior culture. The greatest Hellenistic thinkers in the sciences, particularly Strabo, Posidonius, Galen, and Ptolemy, all wrote in Greek. None of them, however, were from Greece: Strabo was from Pontus, northern Turkey; Posidonius from Apamea in Syria; Galen from Pergamum; and Ptolemy from either Macedonia or Egypt. It doesn't appear that any of these men, who traveled widely, ever visited Athens, the classical nexus. They were all from outlying areas. Their contributions provided the basis for study in medicine (Galen), geography (Strabo, Posidonius, Ptolemy), natural history (Posidonius, Strabo), and mathematical astronomy (Ptolemy, Posidonius) for over a thousand years thereafter—though not in Europe.[7]

Alexander's victories brought major linguistic change to the region. For centuries, Aramaic had acted as the lingua franca of Asia Minor and Egypt, the language of the Achaemenid Empire (522–330 BCE). Though divided into a series of dialects, Aramaic in written form was highly standardized and employed as an official medium for diplomacy, trade, religion, and scholarship across the entire Persian imperial realm.[8] The Aramaic script was even adopted by other tongues as orthography during this time. Alexander's conquests brought all this to an end, spreading Greek across the region. In most areas, Attic-Ionic Greek became the new lingua franca of government and commerce and was learned as a second language to local tongues, which generally were not displaced (Persian, for example, remained a regional tongue in Iran and surrounding areas). A standard written form of Greek replaced Aramaic as a formal, regional medium for political administration and, to some degree, trade.[9]

After Alexander's death, his conquered lands fractured into several parts: one to the east encompassed by the Seleucid Empire, reaching from the Levant to Pakistan; one to the north known as the Greco-Bactrian Kingdom; and Egypt. In the first two of these, Aramaic continued to be used alongside Greek, though in reduced fashion. By the second century CE, Greek had been largely replaced in Persia, where it had never penetrated very deeply, by Pahlavi (Middle Persian) and to the north, in Central Asia, by Khwarezmian, Sogdian, and Bactrian, which persisted till the Islamic era.[10] In Egypt, things remained Greek. One of Alexander's generals, Ptolemy, appointed as satrap, built the city of Alexandria into a great commercial and intellectual center, importing Greek thinkers, artists, teachers, and settlers. The Ptolemaic dynasty he established further deepened the implantation of Greek culture. As a result, the Egyptian language, known as Demotic, fell into disfavor and disuse for nearly all functions of administration, commerce, and scholarship, but continued as a vernacular.

During the Hellenistic period, then, the Greek language provided a critical medium for brilliant men from diverse parts of Alexander's empire. Such men had access to a Greek education, by far the most sophisticated and advanced at the time. No less, they had access to the entire body of Greek scientific thought, an unmatched wealth of intellectual content. Due to fully developed trade and commerce in the region, they could travel to different lands in the Greek-speaking East and find in one place a text or teacher that they couldn't in another. They could converse and debate with other thinkers and teachers of the highest caliber. And they could share and discuss their own work, receive criticism, then hone and perfect it through their own teaching. Greek, in other words, bolstered by an actively interconnected commercial-cultural setting, provided a nurturing embrace for both the availability of knowledge and the opportunity and motive for adding to it.

In 330 CE, political power shifted to the East when Constantine moved the seat of the Roman Empire to Byzantium. Latin was then officially established as the language of the court and the newly installed Christian church. But as before, Greek remained the tongue of scholarship, the arts, trade, religion, and the sciences. This was all the more true after the fall of Rome n the late fifth century CE. Greek continued as the language of teaching in the Eastern Mediterranean region and took on a specific preservative role, as progress in science largely ceased in favor of copying the writings of previous thinkers from Aristotle to Ptolemy. What happened to Greek science thereafter is an interesting chapter. Many manuscripts appear to have remained in Constantinople, but were not actively studied due to purges of secular (non-Christian) learning, especially under Justinian. Instead, scholarship was continued by Nestorian and Monophysite teachers, who were forced to migrate eastward in the sixth century into Syria, Iraq, Arabia, and Persia. Within the limits of the Sasanid (Persian) Empire, the Nestorian and Monophysite teachers took up the Syriac (Aramaic) language that served as lingua franca there and translated some of the works of Greek science into their new tongue in such cities as Edessa, Nisibis, and Jundishapur. Since the death of Ptolemy in the late second century, little new had been added to Greek science; the Nestorians mainly preserved and taught what already existed. Yet it was from these areas, and via Syriac most of all, that this science with centuries of commentary passed into the able hands of the Arabs.[11]

Latin and Science: The Era of Rome and Afterward (Second Century BCE to Tenth Century CE)

Latin, of course, gained the status of a true lingua franca with the expansion of the Roman hegemony in the first century BCE. Gradually, it replaced local tongues as the language of politics, trade, lawmaking, and military matters in the western, central, and northern parts of the empire. In many areas, it also became a vernacular, the tongue of culture as well as administrative power.[12] The Romans were not much interested in Greek science, in its native form. What science they took from the Greeks came from literary sources and from digests and handbooks. A goodly part of Roman astronomy, as understood by the educated elite, was learned from a Greek poem, the *Phaenomena* of Aratus. Handbooks, on the other hand, were popularizations of more theoretical writings, assembled by various authors to fill a demand for study and teaching in the Eastern Roman Empire. Aristotle's pupil, Theophrastus, wrote a number of handbooks later translated into Latin, as did Posidonius (135–51 BCE), known as the greatest polymath of his day. Posidonius's works were used throughout the empire—so copied and plundered were they that nothing remains of them save a few fragments in other authors' works.[13]

Rome's choice of sources for its science proved pivotal. It helped guarantee that the sophisticated portion of Greek thought would remain in the Near East, to the benefit of Islamic thinkers, while Europe would inherit a far simpler brand of knowledge, comparable to a middle-brow encyclopedia. Though it was common in the late Republic for young men of able mind and means to learn Greek and study with Greek tutors, this never led to a taste for Greek science, except via the handbook tradition. This is even true of the most sophisticated scientific work of the late Republic, *De rerum natura* (On Natural Things) by Lucretius, a poem that seeks to "heal" ignorance about the nature of reality. Still more influential was the work by Pliny the Elder (first century), the *Naturalis historia*, a huge compendium of descriptions gathered from older Latin sources on every subject under the sun, from astronomy and mining to magic and makeup. In Pliny, there is no use of mathematics, no diagrams or tables, no proofs. Rather, there are endless recitations, usually from second- and third-hand sources. As the magnum opus of Roman science, the *Naturalis historia* became a standard reference work for the study of the sciences in Europe until the Italian Renaissance.[14]

Roman indifference to Greek science probably stemmed from the role of knowledge in Roman society, how scholarship was organized and rewarded. A fundamental concept was that strong minds should serve the state and the people; knowledge needed a practical aim and was best conveyed in language, not mathematical symbols. If we expand our view to fully include engineering, then, the Romans step onto center stage. Innovations in military technology, building and architecture, hydraulics, construction materials, transport, metalworking, and sanitation, many of them improvements from Etruscan, Greek, and Egyptian models, were put to use in the largest public works projects of the ancient world. Hydraulics was a particular area of invention, begun by the famous *cloaca maxima* ("great drain"), which dried the marshes of early Rome. Such works were matched aboveground by the monumental network of aqueducts. It was this dual capability of conveying away "bad" water and bringing in the "good" that largely allowed Rome to grow as it did, reaching perhaps a million inhabitants by the early empire period.[15]

As a lingua franca, Latin was influenced by scientific terms from the Greek but did not itself become the tongue in which the strongest scientific minds of the time chose to express themselves. Neither Aristotle nor Ptolemy had any place in the curriculum of the Western world until many centuries later. Still, what Latin did achieve was critical, preserving the science that Rome offered. When Martianus Capella in the fifth century set out the seven liberal arts in an ornate poem known as *The Marriage of Mercury and Philologia*, he was following Roman models (grammar, dialectic, and rhetoric in the trivium, geometry, arithmetic, astronomy, and harmony in the quadrivium). That he left out architecture and medicine, which earlier Roman authors including Varro included, should be counted as unfortunate, as Capella became a curriculum guide in Europe until the twelfth century. The point, however, is that Latin made such a curriculum possible.

After the fall of Rome in the late fifth century, Latin underwent a division between the spoken and the written word. Invasion of Italia by Germanic peoples, and long-term contact with other languages in the more distant parts of the empire, had imposed many local changes on spoken Latin. By the seventh century, it was no longer possible for speakers across the former empire to easily understand one another. Early versions of the Romance languages made spoken Latin essentially a dead tongue among the general population. Yet it remained the written lingua franca of intellectual culture throughout Europe and Britain, due to its central role in the Roman Catholic Church. This basic function held for a thousand years thereafter. Until the twelfth-century renaissance and even afterward, the church acted as a preserving institution for what scientific knowledge had been transmitted to Europe. Though written Latin also seems to have undergone an important degree of localization by the Carolingian Renaissance,[16] as judged by the fairly widespread use of certain texts, this didn't at all prevent Latin from remaining a regional tongue in intellectual and religious matters. Astronomy was the science most taught due to its service in helping fix the Christian calendar, including the dates of sacred days and festivals including Easter. Four works were used to study the planets and stars: Martianus Capella's *Marriage of Mercury and Philologia*, Isidore of Seville's *De natura rerum*, Bede's *De natura rerum*, and above all, Pliny's *Naturalis historia*. These works were copied, excerpted, and in Pliny's case, reorganized and even rewritten, many times over. Little or no new material was added to any of them, except for diagrams.

The overwhelming function of Latin during this period, vis-à-vis science, was thus twofold: preservation and circulation. The written language, though far from fully standardized, remained a performing lingua franca in literature, law, and theology, as well as science. Within the monasteries and abbeys, it was still spoken, though in society generally it was already a dead language for speaking but a living tongue for writing. It allowed scribes to copy and re-create texts, which were then used in the libraries and schools, and it urged authors to produce new works that could be read throughout Europe and bring them fame. It gave scholars and students the ability to travel and interact with their peers throughout Europe, the former to teach in new places or establish new schools, the latter to search out and journey to the best of these. It's often said that scholars of medieval Europe had Latin as a "second language." This was probably true in many cases, but may have been restricted to those who actually taught in the language (not all did). Whatever the case, the preservative and circulatory functions of Latin remained dominant until the twelfth century, when new powers were added to it. Beginning in the 1100s, a new chapter began—the massive translation of Greco-Arabic scientific literature.

Arabic Science: The Great Flowering

With the success of Islam in the seventh century and the stunning expansion of its lands, much of the region conquered by Alexander and later inhabited by the Nestorians came under control of Muslim society. A great flowering of intellectual appetite took place, reaching its zenith during the reigns of al-Mansur, Harun al-Rashid, and al-Mamun in the eighth and ninth centuries. The reasons for this flowering are multiform. They begin with the sociocultural, economic, and political realities of the new empire in its major phase of consolidation. But they also relate to diverse intellectual traditions now brought together under Islam's protective roof. The 'Abbasid court, with its interest in astrology, gave favor to Sabian astronomers from Harran in southern Anatolia, a center of Greek speakers and star worship, and to Persians who had been influenced by Indian astronomy. The 'Abbasids also moved to legitimize their rule, particularly over restive Persia—whose language, Pahlavi, was widespread in the eastern portions of the empire—by claiming they were the true successor to the imperial dynasties of that region, which had already founded a tradition of translating and utilizing works of Greek and Hindu science.[17] No less important were Muslim victories in Central Asia, including that over Samarkand, where Chinese technology, including papermaking, was well established.

The early 'Abbasid caliphs visualized their society as reflecting a divine plan, with science given a decisive role. This vision included a "heavenly city," ripe with spiritual wisdom and worldly knowledge; a city that would collect within its walls and the minds of its scholars the highest achievements of the known world.[18] This city was to be Baghdad, built by al-Mansur as a perfect circle, with the auspicious date for groundbreaking determined from the stars. The result was magnificent: elegant palaces and mosques, broad avenues and squares, lush parks and gardens, rich bazaars, numerous colleges, libraries, and hospitals, and ships of all kinds lining the quays, including those from China. Water was supplied by aqueducts, expertly engineered, and people of many nationalities flowed through the well-designed streets. The people and the army spoke Aramaic, Persian, Sogdian, tongues from Transoxania and Khurasan—but all needed the language of the Qur'an. Education was a priority—"the ink of the scholar's pen is more precious than the blood of a martyr" became a traditional view—and learning in the so-called foreign sciences was accorded much respect.

For the history of science, this period was epochal. It brought the translation into Arabic of hundreds of foundational texts from Greek, Syriac (Aramaic in the Near East), Pahlavi, and possibly Sanskrit as well.

Essentially, the entire corpus of Hellenistic science, plus the philosophy, medicine, astronomy, and mathematics of Persia, came into Islamic culture between the late 700s and early 900s. Islamic intellectual culture showed itself to be the complete counter to Rome, having little interest in literary works but unquenchable thirst for philosophical, mathematical, and scientific thought, from Aristotle to Ptolemy.[19]

Linguistically, this meant the end of both Greek and Syriac (Aramaic) as lingua francas of Greek science in this region. Greek certainly continued within the Byzantine Empire, until its final fall in the fifteenth century. But elsewhere it was gradually replaced, as was Syriac, Coptic, Berber, and a number of other local tongues, by Arabic. This did not happen overnight; Arabic at first was an oral tongue of desert nomads. It took two centuries for this provincial speech to absorb influences from its conquered territories to become a magnificently flexible, expressive, and cosmopolitan instrument.

To the east and north, a different overall story is evident. Central Asia, conquered in the later seventh and eighth centuries, had come to use Pahlavi as a lingua franca among a wide variety of local tongues, such as Sogdian, Khwarezmian, Uyghur, Bactrian. Most of these died out by the thirteenth century, mainly under pressure from Persian—the language that Pahlavi became once it adopted many Arabic borrowings and took on the Arabic script, giving up Aramaic characters. This adoption of the Arabic writing system also happened for other tongues, including Kurdish, Azeri, and Urdu, that continue to be spoken today. Pahlavi itself was repressed for a century or more by the Arab conquerors, but by the ninth century all of Persia had converted to Islam, and loyalty to the ancient and culturally deep-rooted Persian language reasserted itself. Arabic never caught on very well as a daily tongue in the eastern part of the empire, while a highly cultured, Persian-speaking class of officials came to dominate the empire's secretariat. Nonetheless, Arabic commanded the highest authority. It was the language of administration, taxes, military affairs, and religion, and also became the reservoir for science.[20]

The translation movement was a conscious one. It "cut across all lines of religious, sectarian, ethnic, tribal, and linguistic demarcation . . . [its patrons being] Arabs and non-Arabs, Muslims and non-Muslims, Sunnis and Shiites, generals and civilians, merchants and land-owners, etc."[21] The translators, too, were of diverse backgrounds, from many parts of the empire. Greatest among them, known for the elegance of his Arabic, was Hunayn Ibn Ishaq (circa 809–873), a Nestorian physician working mainly from Syriac, who rendered dozens of works by Galen, Aristotle, and others. He was among a number of translators sent by patrons to various parts of Asia Minor to search out works of the highest quality and reputation. Some of these works, such as Ptolemy's *Almagest*, were rendered a number of times or edited and corrected by later authors. Translators by the ninth century were also writing their own sophisticated treatises in Arabic. Thabit Ibn Qurra (826–901), for example, from Harran, was recognized for his linguistic skill in Greek, Syriac, and Arabic, and was brought as a young man to Baghdad as not only translator but educator and court intellectual. His writings indicate that a community of scholars had grown up in the city to include Muslims, Sabians, Christians, Jews, and Zoroastrians, all working and living in close contact to create the best material of the "foreign sciences."[22] By the late tenth century, the library of texts was fairly complete and the translation movement came to an end, having essentially run out of material.

Another reason for movement's end was the growing number of original works by scholars already in Arabic. These authors improved, critiqued, updated, and transcended the translations. Al-Zarqali, for example, from southern Spain, corrected Ptolemy's geography and helped produce the famous *Tables of Toledo*, containing the most accurate star positions for centuries thereafter. Al-Khwarizmi, who hailed from modern-day Uzbekistan, conceived of the first systematic methods for solving linear and quadratic equations, methods that served as the basis of algebra (the word itself is derived from his book, *Al-Kitab al-mukhtasar fi hisab al-jabr wa-l-muqabala* [Compendium on Methods of Calculation by Completion and Balancing] and the term *algorithm* is the Latin version of his name). It was another of his works, on Hindu numerals, that introduced the modern numerical system (that we know, ironically, as "Arabic numerals"). There were many other such authors: Ibn Sina (Avicenna) and al-Razi (Rhazes) in medicine; al-Battani and al-Biruni in astronomy and mathematics; Ibn al-Haytham (Alhazen) in physics, mathematics, and optics; al-Kindi, al-Farabi, and Averroes in Aristotelian philosophy, all authors from many parts of the empire: Central Asia, Persia, Mesopotamia, Andalusia.[23]

Thus, the expansion of scientific thought was fully commensurate with the rise of Islam and the Arabic language as a new lingua franca of vast extent. By the 800s, Arabic was the official tongue in lands that reached from the borders of China to the Atlantic Ocean. This did not last in practical terms; as noted, by the ninth century Pahlavi grew to be the true lingua franca of the Eastern Roman Empire, including much of India. Over time, Arabic would interact with local tongues across North Africa and the Near East to create a diversity of evolving vernaculars. Well before the modern era, these had become unintelligible to one another, similar to the Romance languages that evolved from Latin. Written classical Arabic remained the textual lingua franca for the central and western regions of Islam.[24] As with Latin, it was learned in schools, in formal settings, but became extinct as a vernacular.

In its rise, classical Arabic eliminated other tongues, even as some of these acted on it, dividing it. Written Arabic provided a medium for students, teachers, and scholars to have access to an enormous body of writings and to contribute to this corpus in turn. As a language of conquest, of political, military, economic, and eventually intellectual power, it displaced Greek and Syriac and a number of other tongues, just as classical Latin had done before it. Its magnanimous extent destroyed it as a spoken tongue for all the Islamic world, while making it a formidable language of written knowledge across a vast region. Works by its translators and authors came to populate the great libraries assembled at Baghdad and Toledo, places separated by over twelve hundred miles and a dozen earlier cultures.

Latin in Europe: From the Twelfth-Century Renaissance to the Scientific Revolution

Just at the time when interest in the sciences was beginning to decline in Islam, it rose in Europe, again as part of profound changes in social and technological conditions. Between the eleventh and thirteenth centuries, new contacts were forged between the European mainland and Islam. From the Crusades, Europeans gained firsthand acquaintance with Islamic society and, eventually, the Arabic language. The late eleventh century saw the Norman conquest of southern Italy and Sicily, areas that had seen a succession of rule by the Lombards, Byzantines, and most recently Muslims, creating a diverse, multilingual society. In 1085, after uniting most of northern Spain, Alfonso VI of Castile took possession of Toledo; Arabic speakers were now living under Latin rule. Toledo opened its fabulous libraries, offering nearly the whole of Greco-Arabic science, while collections in both Arabic and Greek were discovered in Sicily. By the 1140s, a new era of translation had begun, pursued by bold individuals working for no central authority but for fame and posterity.

The new translation movement arose as part of an awakening in many areas of society. A time of peace settled in northern Europe after the end of the Viking invasions, aiding growth in commerce. Agriculture benefited from the warming of the climate that peaked around 1100, and from innovations in waterpower and the wheeled plow. Mechanization entered many industries, such as tanning, textiles, and mining, which helped expand new trade with the Eastern Mediterranean region. With less war and more food, population surged in towns and cities, where more building occurred than had been seen since Roman times.[25] Like Islam in its early period, Europe now embarked upon an effort to create a worldly society. This meant new migrations—liquid capital, the widespread transfer of goods, the movement of artisans, teachers, and students, and the circulation of books and ideas.

The greatest transformation, indeed, took place at the level of mind. The aptly named twelfth-century renaissance affected nearly every domain of intellectual endeavor, from literature, as in the troubadours and Arthurian romances, to architecture, with the building of the first great cathedrals. There may be no better mark of the new place of learning in a Christian world than the west façade of Chartres Cathedral (sculpted circa 1150), where the seven liberal arts surround the signs of the zodiac, which in turn envelop a seated Virgin Mary with the infant Christ. From the growth of schools there developed the first universities in the West, a new institution to absorb the new works of Greco-Arabic learning.

The translation movement occurred within, and contributed to, this restless context. Its main phase spanned about 1130 to 1250, with activity first centered in Spain, on works in Arabic and, to a lesser extent, Hebrew. Translators here included the likes of Adelard of Bath, John of Seville, Hugh of Santalla, Plato of Tivoli, and above all Gerard of Cremona, the Latin equivalent to Hunayn Ibn Ishaq. In the later twelfth and thirteenth centuries, the centers of work shifted to Italy and Sicily and to Greek manuscripts, rendered by the likes of William of Moerbeke. By 1250, the entire library of Greco-Arabic science had been brought into Latin, plus works of Arabic literature, rhetoric, and the Qur'an itself.[26]

It is, in fact, interesting to contemplate the parallels between this episode and what had occurred in Islam four centuries earlier. The scale of the work and choice of texts are very similar, as is the translators' awareness of their historical role. A major difference is that translators from the Islamic world were well supported by patrons of every kind, while Adelard and Gerard were not. But underlying a great deal of what occurred in both cases was a new culture of the text, including changes in the technology of writing—better-quality pens and ink, the introduction of paper in Islamic lands—and the great value given written documents.[27] Both the Islamic world and Europe saw a large increase in the number of authors, the extent of literacy, and the number of readers, all using the same lingua franca.

The era of translation altered the Latin language, but less so than Arabic, which had been entirely oral. Though spoken within the walls of monasteries and churches, Latin was school learning, preserved not by the infant's ear and mother's voice but by long years of classroom study. Being dead gave it a strategic aspect: it allowed each new generation to easily read and comment on works of the past. Yet the late medieval period brought a great expansion in this book-learned language even as there appeared the first full works in vernacular languages including Catalan, German, French (Provençal), and Italian. The European vernaculars veritably burst on the scene after 1100 and increasingly took over the arts. This further enthroned Latin as the tongue of highest learning and the church, making it seem timeless even as it gained new aspects. Vocabulary for the new sciences, after all, did not yet exist and had to be calqued or invented. Invention came from a near orgy of neologisms for Greek and Arabic terms. What eventually emerged was scholastic Latin, more flexible and flowery, capable of abstruse expression, but able to contain and teach what had come from the libraries of the Islamic world.[28]

Latin thus remained a lingua franca by both artificial and creative means. More than ever, it was the vehicle that allowed teachers, students, and pilgrims to migrate across Europe. If pronunciation varied between England and Italy, mutual intelligibility of writing and reading was never a problem. That it remained artificial was apparent. It was rarely used by women or children, for example. As the lingua franca of learning, Latin interacted far more with nonnative tongues in Europe—Arabic, Greek, Hebrew—than native ones. In its relations, it had far more commerce with other textual lingua francas than with the local languages to which it had given birth.

How did all this play out in the centuries that followed? Scholastic Latin flourished into the 1400s, when the Italian Renaissance brought a shift toward what we would call the humanities, away from the sciences and theology of the prior era. Science was not ignored, far from it. Yet major advances would have to wait until the mid-sixteenth century, for Copernicus in astronomy and Vesalius in medicine, for instance. By this time, Latin had undergone a mutation. Influential humanists, such as Petrarch, claimed a need to return to the purest forms of the language, identified as Roman writers of the late Republic, especially Cicero, Virgil, and Horace. Works by these authors were viewed as exhibiting the best Latin that had ever been achieved. This meant refinement and latitude for eloquence. But it also meant that school Latin mutated into museum Latin, a tongue restored from the distant past, even more removed from daily life and more difficult to learn. Museum Latin was further empowered by the appearance of the printing press and movable type.[29]

The new mass production of books made knowledge more affordable and textbooks more accessible. For a time, it helped maintain the dominance of Latin in classrooms across the Continent. In France, Germany, and Switzerland, children were not offered any instruction at all in their native tongue until the late 1500s or well after.[30] Even the Reformation, when it swept much of Europe after 1540, giving new privilege to national tongues as a means to combat the Catholic Church's hold on ritual, did little at first to displace Latin from the classroom. An observer gazing over Europe's universities in the first decade after 1600 might be forgiven for believing that Latin had such an influence in the sciences that it would remain in place forever, if not longer. Even Francis Bacon, herald of the new "experimental philosophy" and a powerful voice in the early 1600s for doing away with Ciceronian finery, chose Latin for most of his works.

But Bacon, with Galileo and Descartes, also helped forge a new standard. Growing nationalist feeling in Europe, aided by the rise of French power and the Anglo-Spanish conflict (1585–1604) as well as the early phases of colonization, had a major impact on use of vernaculars. Bacon and Galileo helped set the model in science initially, putting their works in Latin and in their native English and Italian, respectively: Bacon for his *New Atlantis* (1626), Galileo for his most famous work, *Dialogue concerning the Two Chief World Systems* (1632). Galileo even explained his choice, noting he chose Italian because "I wish everyone to be able to read what I say"—"everyone" now indicating his countrymen, who might be his students or patrons, not the scientific world of Europe. Descartes employed Latin until the 1630s, but then published *Discours de la méthode* and *La Géométrie* (both in 1637) in French. Equally important, Christian Huygens also composed his early works in Latin but by the 1680s began to employ Dutch and French as well. Van Leeuwenhoek, a draper having no knowledge of any language but Dutch, gained fame in the 1680s through his letters to the Royal Society of London, which had them translated into English; their importance was viewed as so high by Robert Hooke that he learned sufficient Dutch to read them. The English, in particular, were the forerunners of Latin's demise—Hooke, Robert Boyle, Edmond Halley, Christopher Wren, and Newton all published work in the vernacular, though they also used Latin on occasion. Yet these men showed—indeed, they established—that Latin was no longer required for science in Europe.

Latin was thus gradually stripped of its power. One area of dominance after another was abandoned. Its use in commerce had been the first to go, as early as the thirteenth century, then government and politics during the Renaissance, followed by literature and the arts by the late 1500s (the great authors Rabelais, Montaigne, Cervantes, Tasso, Spenser, and Hakluyt all wrote in their national tongue), and its replacement by French as the lingua franca of diplomacy. Science, medicine, and mathematics held on to Latin the longest, partly due to frequent transnational contact. The language did continue as the voice of higher learning into the eighteenth century. Universities continued to conduct classes in Latin and often encouraged students to speak to each other with this language. But even here, within the walls of the ivory tower, change soon came. Francis Hutcheson, famed mentor of Adam Smith at Glasgow University, was only the first to forsake Latin for his lectures, starting in 1729. By this time, however, the literary world had already moved far in this direction. Records show that in 1700, new books published in Latin accounted for roughly 50% of all titles in Italy, 38% in German-speaking countries, less than 25% in France, and under 10% in England, center of the Scientific and Industrial Revolutions. By 1800, even in the German-speaking world, the figure fell below 5%.[31]

As a lingua franca, Latin therefore died twice. As a language of the people, it vanished in the early medieval period; its survival thereafter was never false but confined to elites, by artificial, socially prescribed means. But these, too, proved mortal. Over time, there was a decline from church Latin to scholastic Latin, then to museum Latin, and finally to archeological Latin. Reduced to a skill presumed to provide direct access to "the best that has been thought," in Mathew Arnold's famous phase, the tongue of the Romans could not be sustained as a medium for all in the new era of nationalism.

China and Science

The situation of the Chinese language and science begins from a foundation similar to that of Arabic. Chinese written script came to be adapted over time as orthography for a variety of other languages. This didn't happen only for distinct tongues within the borders of what we call China today (even now, *Chinese* has two meanings for linguists—one referring to a single writing system, the other to an entire family of speech forms).[32] *Hanzi*, or Chinese characters, also came to be a regional orthography for languages in surrounding countries that did not yet have any writing system of their own: Korean, Japanese, Vietnamese, Mongolian (in part), and possibly a number of languages in Southeast Asia as well. Korea and Japan adopted ideographic writing between roughly the third and fifth centuries CE. For twelve hundred years thereafter, the Chinese written language served as the tongue of scholarship, including science, throughout East Asia, even as these countries evolved their own separate orthographies.

As for Chinese science, there was an especially long period, or series of partially linked episodes, during which the study of nature and advances in technological innovation reached remarkable levels. No small portion of these advances, particularly in technology, arrived in Europe via Islam, thus in many cases through materials and texts brought by Persian and Arabic speakers in the tenth through fifteenth centuries. But well before this, scientific thought in China had established itself in much of the East Asian region.[33]

Indications of this can be derived, for example, from the origins of science in Japan, starting with intellectual relations between China and Korea in the period of the Warring States (475–221 BCE) and spreading to Japan in the Han dynasty era (206 BCE to 220 CE).[34] Chinese colonial settlements in Korea brought technology, writing, and literacy. Southern Korea acted as a gateway: requests by the Japanese emperor for teachers of medicine and calendar-making (astronomy-related knowledge) were made as early as the sixth century, with contact growing into the T'ang dynasty (618–906 CE), by which time literacy in Chinese was established among the Korean and Japanese elite.[35] China in this period, to ensure its internal stability, was actively spreading its own cultural and administrative models. From the other side, Japan sent large-scale missions to mainland China, sometimes with several or more ships and dozens of technical people who would remain for a year or more and return with loads of books.[36] These missions ceased by the late ninth century, yet their impact had been made. Much of Chinese science, at a rudimentary level, and a selection of its technology (for example the abacus, papermaking) had been transferred. For many centuries thereafter, China remained the unchallenged authority for knowledge of medicine, the heavens, and the earth. As late as the sixteenth century, almost no works in these areas had been written by the Japanese, who continued to use only copies of original Chinese books (though with commentary and clarifications).[37] This suggests the depth of authority granted to such works.

It was during the Song dynasty (960–1279 CE), however, that China reached the apex of its own scientific achievements. This era marked a distinct rupture from the past. A long tradition of fractious, hereditary aristocracy destroyed by civil war was replaced by a vast bureaucracy of scholar-officials grounded in Confucianism. Education, literacy, and authorship became paths for advancement. Song intellectual culture had two basic threads: first, to selectively revive classical learning by revitalizing the best that had been thought and invented; and second, to create new syntheses of knowledge and, in a number of cases, to extend these well beyond any precedent in a fully exploratory way.[38] In the case of medicine, for example, the government was directly involved in locating, revising, printing, and supporting the teaching of written material from the T'ang and Han dynasties, with resulting manuscripts transported to all administrative parts of the country, even the most distant, from which they had an impact on many parts of East Asia.[39]

The Song era brought relative peace and unity over much of the empire, allowing for improved travel and communications, with much trade beyond borders as well as exchanges of knowledge. The civil service was greatly expanded, systematized, and established as the goal of the educated elite by a universal examination system. This meant a general channeling of scientific and technological efforts toward uses that could serve the state. Yet the Song period was also a time of remarkable polymaths such as Shen Kua and Su Song (eleventh century), who investigated on their own a striking array of natural phenomena, from those related to geology, astronomy, and mathematics to botany, zoology, and metallurgy.[40] Still, the dominant advances, invented by the educated elite, were practical in nature: gunpowder, the compass, movable type, paper money, improvements to textile weaving, canal-lock systems, ship design, herbal medicines, and metallurgy. All these inventions spread rapidly due to advent of woodblock printing. Agriculture, too, was transformed into a powerhouse of economic prosperity because of the extensive adoption of wet-rice cultivation. Treatises and manuals on all these new technologies were printed and circulated to many districts.[41] Such texts were often simply written and illustrated, greatly increasing not only their utility but their popularity as well. As the same methods of rice paddy cultivation appeared in Korea and Japan at about this time, the transfer of Chinese agricultural manuals to these countries seems likely.[42] This is also suggested by the outward-looking aspect to official Song China, which had a great interest in geography and cartography.[43]

The Mongols conquered China and Korea in the thirteenth century, due in part to the appropriation of Song technology. Mongol arrow- and spearheads changed from bone to iron; inflated skins and log rafts were traded for a true navy; and use of armor and siege warfare advanced to a level at which no Chinese city could stand. In each case, new technology was transferred when Chinese experts or commanders were captured or turned to the Mongol cause. Once installed, the new Mongol dynasty (Yuan, 1271–1368) did not impose its language on China with any real success, and it did not adopt Chinese either, though it did continue many institutions (this was a pattern seen elsewhere, for example in the Khanate portion of the empire, where the Mongols adopted Sunni Islam and eventually the Persian language as well). Kublai Khan tried a unique experiment in commissioning a new orthography (square or Phags-pa script) that would express many of the languages within the greater empire. But despite many edicts and laws commanding use of this script, the remarkable attempt proved a failure. Chinese, after all, was already functioning as just such a textual lingua franca for much of East Asia.

Trade broadened in goods and books, and the court became a scene of unprecedented cosmopolitanism, supporting Muslim, European, Tibetan, and Hindu scholars. Yet the main flow of science and technology was westward, through direct contact with the Islamic world and Europe. Papermaking, the compass, gunpowder, and advanced canal-building were introduced to the West at this time. Books of Greco-Arabic science were brought into China, probably through Persia, but had little long-term influence. Overall, the Yuan and succeeding Ming and early Qing periods all demonstrate a decline in the level of scientific discovery and innovation compared to the Song era.[44]

This does not at all mean a stagnancy or falloff in related effort. On the contrary, the stable and prosperous period of the Ming dynasty saw a number of important collections of science-related knowledge. Examples include the writings of Li Shizen and Song Yingxing, who produced large compendia covering the methods and technology of medicine and agriculture, respectively. No less significant were the efforts of Xu Guangqi, who wrote treatises on astronomy, agriculture, and military science, and who collaborated with Italian Jesuit scholar and missionary Matteo Ricci on the first Chinese edition of Euclid's *Elements*, printed in 1607. Also during the later part of the Ming period, the traveler Xu Xiake spent three decades touring all of China, documenting and cataloging thousands of its geographic and geological features. As these examples suggest, a great deal of the science performed at this time involved collecting, sorting, explaining, and also illustrating knowledge that already existed, though in dispersed forms. Advancements in print technology enabled these books to be longer, more complete, and cheaper than before. They were also frequently graced with many diagrams and pictures, enhancing their appearance and usefulness.

These aspects, then, did not in the least weaken Chinese as a lingua franca for related knowledge in East Asia. Up to the nineteenth century, nearly all authoritative works on astronomy, mathematics, and especially medicine continued to be written in Chinese. Moreover, in Korea and Japan, these works were now composed or compiled by native scholars, routinely educated in the Chinese classics. In Japan, the situation was greatly strengthened in the sixteenth and seventeenth centuries by a turn toward neo-Confucianism, promoted in no small way by the Tokugawa government, which sought to adapt the Chinese model of imperial bureaucratic control.[45]

Written Chinese thus kept its hold for over a thousand years. This was especially true in those nations where advanced Western science lives today—Japan, Korea, Taiwan. When this science first made its appearance in those countries, some striking ironies emerged. European astronomy came to Japan during the 1600s in Chinese—either written by Jesuit missionaries such as Matteo Ricci or as Chinese translations of works by Western authors.[46] Even as late as the 1830s, it was common for Japanese scholars to translate scientific works not into Japanese but classical Chinese, which still served as the tongue of higher knowledge.

Addendum: The Modern Era

The decline of Latin followed closely on the early rise of the nation-state and the expanded use by scholars of the vernaculars, especially French, Italian, and English, but also German and Dutch to a certain extent. As mentioned earlier, influential men of science, including Bacon, Galileo, Descartes, Newton, and Leibniz, all used their native tongue with increasing frequency for their writings during the seventeenth and early eighteenth centuries. By the 1770s, it was legitimate for a new journal editor such as François Rozier, with the larger scientific community in mind, to complain that the use of national languages meant that two or more scholars might be writing on the same problem and never realize it. Rozier, as we've seen, called for scientists to again employ a shared language, in this case his own (French). His plea tells us that whatever its extent in learning elsewhere, French was not being used as a true lingua franca of science. Its reach, far greater than any other European tongue at the time, was not universal the way Latin, Chinese, Arabic, and Greek had been in earlier periods.

We are thus tempted to ask: how did science progress for two full centuries without a lingua franca? Or, to put it a bit differently but more precisely, how was important work in one language transmitted to speakers of other languages? The simple answer is that a fair number of researchers during the eighteenth, nineteenth, and early twentieth centuries were multilingual and could therefore read and translate the scientific literature of other countries. For the majority who were not so linguistically capable, these multilingual workers became mediators and did the hard labor of transmission themselves. This labor took three main forms: (1) translation of papers, speeches, monographs, and books, done by well-known (even famous) scientists as well as lesser-known ones; (2) various types of reporting, such as scientists in England writing summary pieces on meetings, conferences, or laboratory visits they attended on the Continent; and (3) critical reviews of the literature in other tongues, meaning actual review articles as well as book reviews, discussions of an individual scientist's recent or overall contribution, obituaries of the famous, and more.

We see a number of these forms of transmission in the first issue of the journal *Nature*, dated November 4, 1869. The lead article, in fact, is itself a translation: "Nature: Aphorisms by Goethe," rendered by none other than Thomas Huxley. There is a signed review of *Antiquités préhistorique du Danemarck* (by John Lubbock); four shorter reviews of books in German on biological and astronomical subjects (showing the importance of this language and of German science generally by this time), signed only with initials; two shorter, unsigned reviews of other German works; and a "news" report of significant length by the well-known Scottish geologist Archibold Geikie on the 43rd Annual Meeting of German Naturalists and Physicians. A section of the journal entitled "News" provides a report on an astronomical congress in Vienna and another on the October 25 meeting of the French Acadèmie des sciences, but also discusses recent advances in chemical synthesis published in French, German, and Italian journals (*Annales de Chimie et de Physique*, *Zeitschrift für Chemie*, *Giornale di Scienze di Palermo*).

Most or all of these international contents can also be found in other nations' periodicals, including many from Germany. The example discussed in chapter 3, *Geologische Rundschau*, is excellent evidence of this—with one difference. Journals like *Nature* sought to connect English-speaking readers with the greater world of science, but not to include this world directly. Unlike periodicals such as *Geologische Rundschau*, they did not ask for or publish work in other languages. They continued to rely on multilingual mediators into the twentieth century, until the period when English itself began to serve as the new lingua franca, thus the direct carrier of international content. Not long after, by the 1990s or early 2000s at the latest, journals that had earlier on encouraged submissions in several languages had switched over to an English-only format. Researchers had become, therefore, the linguistic mediators of their own work.

Modern, Western science moved ahead without a true, universal lingua franca for three hundred years, because it had editors, scientist-translators, and others able to serve as the media of internationalism. Above all, translation acted as the key process by which investigators gained access to one another's work across linguistic boundaries. Student mobility, especially involving those from America, Britain, and northern Europe studying in Germany between 1870 and 1914, also played an important part. But translation, in a number of forms—the rendering of papers and books, the summarizing and reviewing of same, the reporting on meetings in other countries, and so on—was the constant during the whole period between the fall of Latin and the rise of English.

Themes and Meanings

Reflecting on the overall history of languages in science, we find that a number of themes rise to the surface. Among the most important would be the following:

- Lingua francas in the past were often imposed by conquest, yet remained in place for a very long time, far outlasting the power and prestige of the original conquerors. This was especially true when the language became integrated into the intellectual traditions of different cultures. Thus, the power and prestige of a lingua franca's native country have been most relevant during the early phases of the language's emplacement.
- As they spread, these lingua francas were gradually nativized to local settings—the further they extended, the more they tended to "break apart" into dialects and, eventually, new tongues. By contrast, their writing systems usually remained intact and persevered as international tongues of scholarship. The writing systems provided the essential possibility for knowledge to be shared, gathered, taught, and advanced.
- Lingua francas have had varied impacts on local languages. They

have led many linguistic groups to abandon their native tongues. Yet they have also caused groups to build new intellectual communities. Whereas Latin helped bring about the extinction of Etruscan, Chinese proved an originating force for early Korean and Japanese science.

- There is little evidence that learning an international language created enormous difficulties in the past or retarded the advancement of science itself. Those authors whose works remain to us do not often, or ever, testify to such problems (though they do complain at times about teaching methods). Multilingualism among scholars was common in many settings, especially urban ones, and knowing the language of power was viewed as an asset, especially by intellectuals. In broad terms, learning Latin, Arabic, or Chinese was a part of scholarly education, as indispensable in scientific areas as in literature or philosophy.
- At the same time, the lack of widespread evidence for difficulty in learning such languages is not at all proof that such difficulty was wholly absent. Almost inevitably there were hurdles, even formidable ones, in certain circumstances, as implied for example by the long time it took brilliant men like Adelard of Bath and Gerard of Cremona to learn a language like Arabic (which they may or may not have ever truly mastered). Moreover, access to such a language was usually restricted to certain classes of society, and major linguistic groups did resist replacing their own tongue in the vernacular with a lingua franca.
- The sciences have benefited enormously and in varied fashion from lingua francas. Such tongues have helped stimulate and connect intellectual hubs among different societies, urban centers, and even regions, linking them to core areas of power and patronage. Such languages have thereby expanded science by opening it up to the pollinating influences of knowledge from other cultures.
- Another major advantage to science has been the power of a lingua franca to mobilize individual thinkers from different geographic and cultural sites. The ability of scholars to move and circulate relatively freely, participating in a greater geographic-linguistic domain as knowledge producers and teachers, has been an important element in the advancement and spread of scientific knowledge.

- Lingua francas that spread most widely, Arabic in particular, served as unparalleled mediums by which native knowledge could be preserved and used to fertilize science. The tradition of star worship in Harran is but a single example.
- Latin in European science doesn't offer a good analog to English today. For most of its life, Latin was a dead tongue, available only to a tiny elite, thereby limiting access to science for many centuries. These aspects are entirely countered by the realities of English, which is now spoken in living form by over a billion people, taught to many more, and acts as a medium for science worldwide.
- Translation has played a critical role in the advancement of science, both in the transfer of scientific material between different lingua francas and as a process for sharing knowledge when a lingua franca has been absent. This role has involved not merely moving text, but revising and reorganizing and updating it as well. Translation has proved itself a dynamic, multifunctioning process in the progressive expansion of scientific knowledge.
- Lingua francas have helped create elite cultures of the word, no less for science than for literature. Yet they have also done the opposite—built paths for those of ability to enter into intellectual enterprise from lower social positions. The degree to which this occurred differed greatly among societies of the past. The openness of the Eastern Mediterranean region in Hellenistic times and of early Islam encouraged many scholars who were not from wealthy families to take up careers in scholarly subject areas.
- Lingua francas show, beyond any doubt, that science has never belonged to any one nation, culture, class, or region. Rather, it has had an international, even global dimension for a very long time, and shared languages have been a primary reason for this.

Taking these themes together, we see that lingua francas in science have brought both constructive and destructive effects. Yet by almost any measure, the positives have vastly outweighed the negatives. International languages have been a key factor in scientific progress, not least because of the increased mobilization of knowledge they have conveyed into new contexts, where such knowledge has been taken up and expanded by many peoples. As a form of power, these lingua francas have often been restricted to scholarly elites. Yet over time, such elites have proved to be porous, nonexclusive, far from absolute in their use and control of such languages. As revealed by the life and career of Adelard, who mastered two lingua francas, one through an established system of education and the other through his own noninstitutional efforts, the ultimate power of a lingua franca is to overflow any one setting—scholars seeking out other scholars and their texts, beyond the bounds of political and even cultural fences.

Certainly, installing a lingua franca has often depended on force more than favor. Conquered peoples have endured periods of struggle and adaptation. Nations and communities have shown themselves far more willing to accept an international tongue with limited functions, including scholarly ones, rather than a wholesale annihilation of their own speech. A language of power, however, also creates a market, a trade in skills and status. Teachers of Greek to the sons of Roman patricians or the builders of Latin schools in twelfth-century France were responding to, and also helping further, a demand that was at once intellectual, professional, and symbolic. Anyone who had studied under the great polymath Posidonius on Rhodes or at the cathedral schools of Laon or Chartres acquired pedigree as well as preparation and could then become teachers themselves. Such is another way that lingua francas in science have expanded.

For science specifically, one strong message from the past is that lingua francas have through history provided not just gateways but pathways. A regionally shared language, again and again, has been a mode of access to intellectual opportunity, contribution, fertilization, and circulation. Far from acting in solely hegemonic fashion, such languages have helped to open scientific endeavor to new voices, new ingredients, and new possibilities. Of course, other elements in a society needed to be in place for such openness to happen—a tolerant and cosmopolitan outlook, a relatively fluid social structure, a commitment to the value of learning, for example.

When we consider the extent of Arabic in particular and the time when it ruled the sciences in their most advanced form, it seems legitimate to call it a true "world language" of science. With the exception of late Tang and early Song China, Arabic was used during the ninth through eleventh centuries by a higher proportion of those active in (what we would today call) the sciences than perhaps even English in 2013. Actual numbers will never tell, since we do not have the data, either then or now, and it seems unlikely we ever will. But it seems a reasonable guess. Thus, indications derived from Arabic's period of dominance may have particular relevance to English in science today.

The Future of English in Science: Some Thoughts and Forecasts

How applicable are all these points to the case of English? As should be clear from the foregoing chapters, they are entirely, even intimately, relevant. What, then, do they suggest about the future of this tongue and its role in science?

First, that the era of English as a lingua franca has just begun. Global it may be in geographic extent and forms of communication, but global throughout the culture of scientific practice it has yet to become, to the point perhaps where possession of it will define who is adequately trained and who is not. As such, it will become entirely ordinary, an unexceptional element in technical training. A side effect of sorts: discussion and debate will shift away from issues of "fairness" and "marginalization" to questions about best practices in English-language learning.

We can also say that the geopolitical standing of the United States will become ever more immaterial to all this. America may well decline over the coming decades in ultimate prestige and even in scientific preeminence (many feel this has already begun), but any such fall from the commanding heights will not, by itself, erode the use of English as a medium of scientific exchange. As with the Greek tongue in the Hellenistic period, long after first the Macedonians and then the Romans had ended the political power and independence of Greece itself, English has become the language for the most advanced forms of knowledge in every field of science, forms that are expanding daily. Rendering into another language the immeasurable corpus that has come to reside in English—which now exceeds by orders of magnitude the combined scientific writings of all previous eras, and whose volume is not merely increasing but accelerating in its growth—would present a monstrous challenge requiring untold resources, labor, and time. Meanwhile, what the hard and especially the soft (cultural) power of the United States has helped achieve, in a historical sense, is to lower what may have been a native resistance to the spread of

English—a love of Anglo-American culture, from movies to music, among young people throughout the world may well have helped advance the acceptance of English when these people entered training for their chosen professions, including science. As put by an engineering grad student from Thailand, studying in Seattle: "My friends and I knew all Beatles' songs when we were twelve. Learning English in high school and college never a problem for us." This type of phenomenon is part of an early, intermediate stage, when the learning of English remains controversial in some quarters. With the passing of only one or two more generations, this stage probably will have ended.

How long will English remain the tongue of science? It is a question that can't be answered. All evidence from the past suggests that the time frame is many decades at minimum, even centuries. Currently, it has no competitors—we have dealt with the mirage of Chinese in this role (see chapters 1 and 3)—and there are at present no others on the horizon. Yet it is always possible that such a competitor will appear at some point. The case of German also wags a finger: it takes many years for an international tongue to be entirely disassociated from its native country; until then, it can be stained and weakened as a result of horrific deeds committed by that country. For this to happen with English, inner-circle countries (discussed in chapter 2) would have to be guilty of crimes against humanity at the highest level. Above all, the United States would have to pursue atrocities against the international community equal to those of Nazi Germany, such as nuclear extermination of entire peoples. We cannot say this is impossible. No one can know the character and extent of conflict in the latter part of the twenty-first century and beyond.

The technology factor also cannot be underestimated. Forms of Internet communication have accelerated considerably the global spread of English, possibly more than is currently assumed. This implies that from now on, it is possible for lingua francas to rise and fall far more rapidly than in the past. The scenario seems right, on the face of it. It doesn't reduce, however, the staggering abundance of scientific information, communication practices, journals, research institutions, and so on that would have to be transferred to a new tongue. A universal translator, invoked for example by the linguist Nicholas Oestler (in his claim that English will fall by about midcentury and become humanity's "last lingua franca"),[47] would remove all such concerns. It would, in fact, make multilingualism itself unnecessary, even wasteful. Or would it? This does not seem so convincing, given evidence in the form of the limited progress realized by translation technology during the past sixty years. Techno-fix solutions to language challenges and problems have an unimpressive history, as shown by the attempts to use first video and then online courses to teach foreign-language fluency.

There are two different problems to think about here—one dealing with technology, the other human relations. Machine translation (MT) has, over the last sixty years, yielded something less than spectacular results. If it has made distinct gains, it remains unable to translate professional writing or speech at anything approaching a usable level of accuracy.[48] Of course, nothing certain can be said about how far this technology may advance in the future. But we should be aware of what MT really signifies—a techno-fix to the natural complexities of human communication. It can be compared to artificial languages, such as Esperanto, which are also engineered for universal understanding but which fail because of their very artificiality, as we noted in chapter 1. For MT in science, the challenges are enormous, given the details of meaning and the precision needed; knowledge of the actual subject is required (as any human translator will testify), for otherwise the simple rendering of words, phrases, and sentences produces a synthetic, inaccurate, and usually unusable content. The irony is that a global lingua franca such as English, when established as part of worldwide scientific training, will likely prove much more efficient and reliable. From the vantage of human relations, meanwhile, MT seems unlikely to "solve" the complexities of communication, since multilingualism isn't viewed as a "problem" to be overcome in most of the world. MT will also not eliminate the attractions and the huge benefits of face-to-face communication—people talking to each other, in the same language, unmediated, sharing knowledge and ideas and experiences with emotional immediacy and even intimacy. Conceivably, a global lingua franca may act to reduce the demand for MT as a universal apparatus (apologies to *Star Wars*' 3CPO notwithstanding).

One thing that can be forecast with some confidence: the future will see a much fuller expansion of modern science in developing nations. Over time, this will help create new centers of scientific work—perhaps especially in applied disciplines—outside the Organisation for Economic Co-operation and Development, in Latin America, Africa, and Southeast

Asia, globalizing science in a more complete sense. English will increasingly provide a pathway for this to happen; indeed, it will prove essential to networks involved in the transfer of researchers, styles of training, data, and more, both to and from areas of the world that previously remained out of the scientific mainstream. An essential aspect here will be the conservation, however partial, of local knowledge about botanical and ecological phenomena, traditional medicines, meteorological events, sustainable agricultural practices, and what is referred to as tribal knowledge or artisan intelligence. A global language for science will help define such knowledge as a frontier of needed study (invisible to the larger scientific community), and supply the medium for sharing results, particularly those derived by local researchers. Scientific English, in other words, will come to serve a preservative function far more than a destructive one.

What, then, about the tension between world Englishes and writing standards in science? If English continues to fragment into an increasing variety of local versions, some of which may well become separate tongues over time, could something similar happen to scientific communication? Based on the example of Arabic, and considering the trend we see today—the movement toward accepting nonstandard forms of writing in scientific English—a few predictions can be made. Whatever the pressures to publish, the highest standard in technical communication is intelligibility. For gatekeepers, this is linked both to a sense of professional mission or integrity and to more pragmatic concerns. Editors and reviewers in particular have created a culture of sensitivity on this point, so that it is well recognized on the one hand that any compromise in intelligibility leads to an actual loss for science and a transgression of readers' needs, and on the other to a possible loss of actual readership and thus a poorer showing on such widely used metrics of value as impact factor and journal usage factor.

With the expansion of modern research worldwide, it is probable that a growing number of journals will begin to accept nonstandard English before long—but only to a point. Outer- and expanding-circle norms in scientific texts will continue to diversify a certain distance away from strict Anglo-American standards. But this distance, judged by linguistic measures, will be quite short. It may even lead, at some point, to reactionary movements by native speakers that will try to set rules for global scientific discourse. All the advantages and benefits attached to a lingua franca—including those related to international collaboration and the reputation of researchers—would be reduced by a localization of this discourse into varieties that partly matched the diversity of world Englishes. There would be little reason, in fact, to employ English for science instead of local tongues.

Moreover, there is the context of scientific publishing to consider. Beginning in the 1980s, this became a global business ever dominated by for-profit corporations (Elsevier, Kluwer, Wiley-Blackwell, Springer, Taylor and Francis, among others) that seek to sell their products to audiences as large and international as possible. These products include not only journals and databases, but reference material, education and training resources, software, and, in medical areas, decision support tools, among other things. Expansion of scholarly research into large portions of the developing world has provided such firms with new markets for the long term. There seems little chance, in other words, that companies would accept losses in the usability of their products, journals included, by allowing editorial standards to become wholly flexible. It is difficult to see the fundamental motives changing, even as publishing models (with evolving commercial arrangements) shift toward purely digital, online modes of access.

The future will not bring, therefore, a grand opening of discourse in science to a panoply of flexibilities. Editors and also authors will resist the call to full localization, with its eventual abandonment of the edict for global intelligibility. In all probability, this will likely mean continued loyalty to Anglo-American standards for some time, at least at a basic level, and with important variations that will progressively be viewed as international norms, not those of particular nations. These standards will define the most globally recognized, understandable, and (in most scientific minds) reasonable model. The mentioned variations will show that portions of the global scientific community have indeed taken some degree of possession over English; "scientific Englishes" may well emerge in a microscopic sense. But these will not mimic the varieties of "world Englishes" in degrees of separation—a global scientific discourse will suffer no death by compound fracturing.

These are but a few meditations on what past lingua francas suggest about the outlook for English in science. There is no blueprint or template involved in making such forecasts. The past can never be a precise model for the future evolution of language, in any domain; technology, institutions, and intellectual practices are all radically different today from their counterparts in the sixteenth, twelfth, ninth, or second century. Broad patterns have been repeated at most of these periods, however, patterns that have affected very different languages and societies at different points in their development, and consequently seem determined in some fashion by the larger situation of an international tongue. Perhaps foremost among all such patterns is simply that of duration—a full-scale lingua franca of science, encompassing every area of scientific study and writing, has always lasted several centuries or more. For some observers, this might be sobering; it might even call forth dismay or rejection. Yet for scientists everywhere, the reality of having a single, globally reliable tongue in which, like a vast reservoir, generations of the most important data and text will collect and remain accessible will more likely be viewed as a gain, for themselves and those who come after them and thus for science itself.

(6)

Does Science Need a Global Language?

In the long history of humankind . . . those who learned to collaborate and improvise most effectively have prevailed.

CHARLES DARWIN

I have never met Aziz, and wouldn't know him if we sat on the same panel at a geological conference. He is one of the many researchers with whom I have collaborated over the years and, unfortunately, with whom I have never shared a meal or a handshake. In one of his e-mails, he wrote that he'd come from a small town not far from Asyut, a major city along the upper Nile in central Egypt. His father was a shoemaker; his mother looked after Aziz and his three older brothers. His brothers all stopped school and got jobs at the age of fourteen to help pay for food and other necessities, as the family was poor. Aziz, who did well in his classes, was allowed to attend high school and then, surprisingly, Assiut University, a public institution (therefore free). A relative helped persuade his father to allow this, adding that Aziz could live with him in Asyut and not have to pay for housing or board. He also said that because of this education, Aziz would later get a high-paying job that could help support the family and augment his brothers' *mahr* (money given to a bride at her wedding).

Aziz studied geoscience because of a trip he had taken with the same relative to the Red Sea when he was twelve. I see them driving south, first to Qena, along the upper Nile with its pink-and-white cliffs of carbonate swept by blowing sand, then eastward through the great plateau of Tertiary limestones, passing bleached walls in merciless sun, then finally climbing through Precambrian mélange, violently folded sandstone, sheared and crumpled shales, shining masses of green serpentinite. I imagine him nearly dizzy, as every mile brings new shapes and colors. The exposed ledges of land, the broken stones and sand, the dust stinging his face, come from the surrounding rocks, which rise up continually like a frozen, heaving sea.

For his master's degree, Aziz went to Cairo University, where he chose a thesis on Jurassic limestones and dolomites in the Sinai. He was invited to spend two quarters at the University of Florida, studying with a carbonate specialist. Impressed with his work, she advised him to get a PhD and then helped support him with some of her own grant money (she told me he sent a portion of this back home). After earning his degree, Aziz received funding from an oil company to continue his research for a brief time, which he gladly did. He knew he wanted to work in academia and so needed a record of publications. This is where I came in.

Aziz could speak English fairly well and read it excellently, but writing was a wholly different matter. His training in the language had not been consistent or good; only a few of his geology classes had used English, and then not so often. Though his oral skills improved a good deal during his stay in the United States, his written work had not advanced very much beyond learning technical vocabulary. His mentor in Florida spent many laborious hours helping him get his doctoral thesis in comprehensible order. I learned of all this when she contacted me, told me his story, and asked whether, for a reasonable fee, I would edit two papers he wrote. I agreed. The work turned out to be far more difficult than I'd expected—a mixture of translation, interpretation, original composition, and editing all at once. It proved successful, however. Aziz had collected excellent data and made analyses that interested many specialists. On the basis of these papers and strong recommendations, he found a one-year appointment at Al-Minya University. Then, after publication of another article I worked on for him, he was offered a one-year position at a well-known university in Australia (where he was called the "department wog"). One more paper and he escaped this work, ultimately landing at Helwan University back in Cairo, where I believe he obtained tenure. Aziz wrote to ask for my help on some other papers he was preparing. But by that time I had become too busy with my own writing and had to decline. "You are the ghost that keep me alive," he wrote in his return message, a kind of farewell.

Over the next several years, I did see other papers by Aziz in major journals, with various coauthors from the Middle East and Europe. He had told me that he would work on his English every day; I suspected that he wanted to teach in Europe or the United States. After all these years, I may still meet him for the first time. I would like to hear him talk about the Red Sea.

A Reflective Summary

Modern science has experienced nothing short of a linguistic revolution. It is a quiet, bloodless, and evolving revolution, unrecognized as such by a great many but clear enough in its reality and significance. Centuries of dominance by national languages have now been overturned. The resultant fragmentation, even if commensurate with the beginnings of modernism, was a historical anomaly in any case. Now it is ended. A new era has begun.

With this book, I have aimed at a single question about this new era: is it a good thing? It is a question about the present and future of science, about national policies and institutions, about collegiality and success, but most of all about millions of individual researchers, teachers, and students at all stages of their careers. It is therefore a question about people and history as much as it is about knowledge and priorities. It is therefore one that demands evidence, analysis, and, finally, judgment. Before offering an answer, then, a few conclusions are in order.

Whatever position we begin with, the fact remains that English is humanity's first true global language, and the work of science occurs within this embracive context. Yet it is also true that in its globalized extent, English has expanded further in scientific endeavor than perhaps any other domain. While it is used or studied by over 2 billion people in over two-thirds of the world's nations, it is also the medium for over 90% of international scientific communication in every form, throughout the entire globe. Nothing like this has ever existed before. It is a new era for the voice of science.

Science's future is therefore inseparable from English. And English in science is itself inseparable from the advent of digital communications, the rearrangement of the geopolitical order after communism, and the expansion of research throughout major parts of the developing world, where it did not previously exist to any great degree. The new era, then, has a number of dimensions. After more than two centuries of being ruled by competition and cooperation among a small set of Westernized nations, science has passed into the most international period of its evolution. We can safely say in the twenty-first century—indeed, we must say—that modern science belongs to no single group, gender, culture, class, or region. Such is what the global history of science over the past several millennia teaches, and it is what the global language of science teaches today.

* * *

It has taken time for English to be fully and finally chosen by the global scientific community. Favored by the consequences of British colonialism, twentieth-century world wars both hot and cold, and the economic might and postwar eminence of the United States, this choice may appear to have been ordained by nonscientific events. There is no doubt that these events have been fundamental and determining, up to a point. Yet the research community is no linguistic puppet. Strongly affected (indeed penetrated) by political and economic realities, it also has its own priorities, its own internal needs, tensions, and demands. The community's deep faith in the universalism of its knowledge, in the principle of sharing this knowledge with others, and in its power to help benefit human beings everywhere define potent arguments for a communal language. Another such argument comes from the professional demand for global priority claims, for recognition, and for intellectual influence. If national languages, and thus domestic science, have tended to dominate until recently, the push for a unifying form of communication able to rise above all splintering effects has been present since Francis Bacon, in *Novum Organum* (1620), declared the power of words to both corrupt and heal the scientific mind. Now Bacon's native tongue has been given the task of unifier.

At the moment, there looks to be little chance of this changing anytime soon. Even the oldest and most hallowed tradition in scientific taxonomy—the use of Latin as a lingua franca to catalogue new species—has given way before the spread of English.[1] True, we cannot possibly know geopolitical events, and their linguistic impact, even over the next twenty to thirty years. Yet already the volume of scientific information in English is immeasurable. Moreover, it is expanding daily. It includes the largest and most essential databases in every conceivable field as well as more than three decades of research published in tens of thousands of the most widely influential journals—plus conference proceedings, vast numbers of books, monographs, government and corporate reports, and much, much more. The rise of China in international science, along with that of India, Brazil, and other nations, springs from the use of English and constantly adds to it—that is, scientists in these nations are all using the English language in publications, at meetings, in corporate R&D, in student training, and in dozens of other venues, to advance their own careers and disciplines. The world of contemporary science at the international level is pouring out more than 1.5 million research papers each year—more than were published in a dozen languages during the whole of the nineteenth century—and all of them are in English. Many scholars have spoken of the advantages of a global tongue in terms of bringing efficiency to the dissemination of knowledge. But *efficiency* is not the right word. Science is about not merely dissemination but participation, both sharing knowledge and, even before this, providing pathways to create it.

Participation in global scientific activity means using English. Expansion of this one language through all the fields and filaments of technical work in the transnational realm continues unabated. French or Brazilian agronomists, working with farmers in Paraguay on a method to treat invasive fungi, will find Spanish far more helpful than any other tongue. But if they want to publish their results, to provide their data and ideas to the greater scientific community, a different language becomes necessary. This suggests a fertile symbiosis. A global language provides a way for quality national science in any language, whatever the goals of those who fund it, to be shared with the rest of the field, worldwide.

Despite the old saying—a language can be defined as a dialect with a military—languages do not inevitably follow in the patterned shadow of politics. Tying the choice of English to the status and power of the United States is no longer convincing. Nearly every great lingua franca of the past outlived its imperial beginnings, often by many centuries. Moreover, the physical and monetary powers of a particular people have not always dictated the elevation of their tongue above all others. Rome ruled the entire Mediterranean, politically and economically, for half a millennium, yet Greek remained the tongue of knowledge and learning east of the Italian peninsula. The Islamic world after the tenth century broke apart into separate empires, yet classical Arabic held its place in scholarly work for two centuries thereafter. Mongolian did not become the speech of Asia after Genghis Khan conquered its cities; instead, Arabic, Persian, and Chinese continued as regional languages of knowledge.

Fears of English acting the part of deadly monolith or fatal hegemon are misplaced. As one (non-anglophone) linguist has observed, in addition to committing the sin of personification (languages are not killing, conquering entities), such a perception projects onto this tongue all the anxieties, resentments, and blackened opinions associated with the United States and globalization.[2] English is neither bringing cultural diversity to its knees nor spreading linguistic death to all regions and peoples. It is not immanently "American" or "Anglo-American," implanting feverish urges for fast food and shopping malls, wherever it goes on planet earth. Nor, as has been said so often, can the United States possibly control the future of English itself. "A dispassionate look . . . shows that popular discussions hopelessly overemphasize the influence of [America] on the developments of the language as a whole," says the same linguist.[3] English belongs to the world, to all who use it, not to America or England or inner-circle countries as a group. A global language is global by use.

For science, dread at the specter of monolingualism appears unnecessary. National science will not disappear. Some countries may choose English as their own tongue for technical work, but many will not and will continue to function perfectly well, employing English at the international level and their national tongue locally. The greater scientific enterprise will always have multiple linguistic domains.

Today, as in the past, most of the world's people are multilingual. All indications are that this will not change but will even expand. In addition, more than half of all living languages may well disappear in the next seventy-five years or less—a terrible, global loss by any measure. No one should doubt or question whether the hard work to document as many of these tongues as possible, and to save as many as is feasible, should be done. Thankfully, digital technology and the Internet provide powerful new tools to help with this effort. Native tongues are indeed precious and valuable because of the unique oral literatures they have; the knowledge they contain regarding plant and animal communities, weather and climate variation, medicines, and kinship patterns, and the keys they provide to linguistic evolution among nonliterate societies. Is it really legitimate to say that many of these societies have their own forms of science, systematized understandings of natural phenomena and processes? Can oral societies possess types of knowledge that we would be forced to recognize as scientific in some part? Absolutely. There can be no doubt. To deny this truth would be to deny any scientific intelligence to humankind before the invention of writing, despite the development of agriculture, the building of cities, medicine and astronomy and mathematics in early Mesopotamia and Egypt, and so forth. Oral societies have always relied on intimate knowledge of natural phenomena. Loss of these groups' languages would thus be a loss to modern and future science. But there is another reason to save or vitalize as many as possible: these languages are integral to the larger history of humanity and are the only "documents" we have of this history in places where writing never found a use. This alone makes them more than worth saving even if they are spoken by a tiny fraction of the world population.

* * *

Should science have more than one global tongue? Would diversity of this kind, at this level, be an improvement, as some have suggested, opening up technical work to a broader range of conception, method, and expression? There are problems with such a view. Language, in the form of a professional discourse, is never so determining as to wholly command, even imprison, how scientists think, act, plan, and calculate. There is no overarching "Anglo mind" working through the English tongue to profoundly limit what can and can't be done. Moreover, scientific expression, as it appears in international journals, seems in the process of becoming more flexible on its own, in a direct response to the global variability of English itself. We err by assuming that "diversity" defines the moral and intellectual high ground in every case: how would it be simpler, more effective, and more humane for researchers (especially in poor nations) to have two or three languages to decide between, each with limited international reach, rather than one of worldwide extent? And what would these languages be?

Here is an interesting historical matter. In 1900, a talented young scientist from Japan had French, German, and English to choose from for gaining international recognition, with German the standout. Fifty years later, he would have found maximum benefit on the international level in the same three tongues, with English ascendant and a new addition, Russian, now in the mix. A generation later, however, by the 1970s, his options would be altered again. English, Japanese, and Russian have become the major international choices—his native tongue had at last risen to such a level. Yet one more generation, into the mid-2000s, and the scene has again changed: Japanese and Russian are now gone, replaced, perhaps, by Chinese, with English dominant. Thirty or forty years into the future, what choices, besides English, will our eternally youthful researcher face? Meanwhile, what do we say to the molecular biologist from Italy who spent years learning Russian in the 1970s, when the Soviet Union seemed a firmly installed superpower, or the Canadian physicist who learned Japanese in the 1980s, when Japan appeared bound to dominate East Asia for many years? The argument in favor of "diversity," against a global tongue, thus becomes a bet against history. We may think that Chinese will be here generations hence, but the past argues that this is mere assumption. To date, English has been the only language constant at the global level for the last one hundred years.

And this means that English has aided the growth of science in many nations, in no small measure, and in varied ways. It has helped expand the actual, daily diversity of scientific work by opening up possibilities of contribution for young researchers in places as different as Mexico, Turkey, and Vietnam. It has provided a medium for these researchers to access the full knowledge of their chosen fields, to keep current on new findings, and to bring their own work to the attention of their colleagues around the world. It has increased the opportunities for collaboration, for all forms of multinational science, including large-scale projects that advance R&D in otherwise impossible ways, such as the Large Hadron Collider. It has opened corporate R&D to a truly global workforce and encouraged new levels of mobility and circulation for academic researchers, thus more broadly distributing research talent. Not least, it has helped greatly enlarge the international movement of students, who are more able than ever before to acquire higher levels of training. And it has done all this for students and professionals in developing and wealthy nations alike.

This, in fact, is what a global language does—it expands access and participation on many levels. It opens gateways to new degrees of contact and cooperation, entrances to sharing and communication. None of this means that a global language must be a universally democratic one. Nothing guarantees that its benefits *will* happen in every case. A global language for research will mean little in a failed state or one too poor to support scientific training and effort at a significant level. It cannot overcome the effects of political repression, illiteracy, or gender inequality. A global language is in no way a panacea for scientific backwardness or lack of opportunity. Even in countries with rapidly advancing economies and strong intellectual traditions, English may be restricted by realities of income, geography, ethnicity, and more.

Crucial at every step is the teaching of English: how restricted might it be in any specific country, how well is it done, what levels of support does it have, who is doing it, and with what pedagogical goals? These are all critical matters. Much benefit from a global language can be clipped or killed if it is poorly taught, if it is available only to a tiny minority, or if confidence in using it is not helped and encouraged. Where the funds and facilities exist to teach science adequately, there should be the capacity to teach languages, too. But for a significant part of the world, whether due to poverty, politics, corruption, or war, it is not yet possible to do all this, to fully erect a modern scientific culture. It is hoped that these conditions will change. They have improved for many countries in recent decades. Again, there are never certainties. But history is on the side of English.

The language is thus doing what international languages have always done, but exponentially. This includes advantage to mother-tongue speakers and disadvantage to those with insufficient training. As stressed at the beginning of this book, international tongues do create casualties; there is no avoiding this. Linguistic "fairness" doesn't weigh very heavily in people's minds against realities of opportunity and advancement. History suggests that the issue will be a temporary one as competence in the language becomes more routine, but *temporary* could well mean decades. In the meantime, much can be done not only at the level of education and training but also, to some extent, by communication gatekeepers in science. Some ideas will be presented below.

The Answer

So, does science need a global language? Yes, it does. In its current historical phase, it begs for and even demands such a language. A global tongue is needed for science's own growth and excellence. It is required for future progress that ever increasingly depends on international connection, collaboration, and plurality of participation. It also depends on more researchers from diverse parts of the world joining its fold, adding their work and experience to expand the greater community. It is needed in order that all nations have a chance at participating in, and benefiting more fully from, the science, technology, engineering, and mathematics (STEM) enterprise.

Science needs a global tongue for all the many reasons that scientists, engineers, and physicians themselves mention when asked the question—summed up, perhaps, by this comment from a Chinese biomedical researcher: "As scientists, we must learn from each other, write for one another. Maybe this man in a lab in Mexico knows something I must know to understand my own results. How do I find him? In Asia, we know very well what this means. Many of us were isolated for a long time from most of the world. Now we must work to improve our methods and data and make sure we are not serving the wrong masters. We must serve each other." Science needs such a language, then, because researchers need it.

Nations need it, too. Many countries have special sets of challenges with links to STEM: problems of food, water, disease, and electricity in India and parts of Africa; water and food in North Africa; pollution, sanitation, public health in Central America; disease, flooding, resource use in South and Southeast Asia. In most of the countries that make up these regions, a full-scale R&D enterprise spanning every technical field, comparable to that in Japan or the United Kingdom, is unlikely for a long time. More focused and strategic research work, aimed at the more pressing issues, may indeed become the chosen future. It is thus imperative that scientists in these places have access to the knowledge they might need from elsewhere in the world.

There is another point to be made. Formal communication with other scientists, especially through journal papers, is not merely a stage in the social practice of scientific work. It is part of research itself, no less essential and integral than data analysis. Material that fails to be shared is not science; the communal dimension defines it as knowledge, as a contribution. Questions surrounding a global tongue therefore are important in a truly fundamental sense, one that even overflows considerations of culture and identity. A global language for research provides the critical medium for more science to be created and shared by more people, from a greater diversity of settings. A global language does not constrain the sciences but, by its very nature, enlarges them.

We see this in the ideals once expressed by Vannevar Bush for America, ideals of science for national progress that are now global, too. Nations, whatever their level of development, are investing in research centers, education, commercialization. There are several categories of such investments—rich countries that are trying to stay in the global lead in innovation; emerging nations (Brazil, China, India) eager to build full-scale R&D systems to catch up with traditional leaders such as the United States, Germany, and Japan; and developing countries that are most interested in addressing their own pressing, long-term needs. Globally speaking, nations today are looking for STEM solutions to many of the overarching challenges that they face and that humanity faces. If science and technology have become mature enterprises in advanced countries, their greatest future growth will be in the developing world. Scientists in poorer nations wish to raise their own level of achievement by having at their command the greater share of their field's literature and data. Desperately needed, in other words, are connections of access between rich and poor in the geography of world science. A global language is the most immediate and powerful such link. It is an essential means to help defeat the backwardness of isolation.

Lest this sound overly idealistic, we shouldn't believe that a community of researchers defines a model for world harmony and peace. Science is fiercely competitive, often insular, sometimes authoritarian, always dependent on government and corporate support, and therefore never free from political priorities. Yet none of this prevents it from being immensely productive and indeed essential to human welfare. Solving a great many of the world's most pressing and formidable challenges—the very real problems related to growing and providing food; meeting the need for clean water; satisfying the surging demand for energy; reducing disease and stopping future pandemics; dealing with the causes and impacts of climate change; protecting the planet's oceans and biodiversity; and predicting and dealing with future natural disasters—depends directly on scientific advances that take account of specific conditions around the globe, that require the participation of local researchers. It is clear at this point that the scientific *cerebelli* of wealthy nations, with their training and resources, need to partner with those researchers from all other parts of the world in order to solve these deep-seated problems with their many local dimensions. A global tongue not only makes this possible but urges its realization. It is thus not only science itself but the world that needs a global language for technical endeavor.

Can Current Realities Be Improved?

A global lingua franca comes with limits and casualties, as we have seen. My once (and future?) colleague Aziz would agree, but would hesitate to count himself among them. Ironically, his own achievement might well get in the way of admitting his own situation. He is, after all, a clear success story, particularly when compared to some of his colleagues, who (he once told me) wouldn't go to America to study because they were ashamed of their English.

The most serious casualties, in other words, come from unequal distribution of access. Earlier lingua francas of science may indeed imply that these drawbacks will prove temporary, that they represent something of an early phase in the spread of an international language, but will nonetheless require many decades to cure. Knowing this, are there any ways to abbreviate the time frame? I believe there are.

The Publishing World: Some Brief Suggestions

Editors of prominent journals obviously have considerable power as gatekeepers in scientific communication. One of their tasks is commonly understood to be that of maintaining norms in technical writing. Yet like all domains of language use, scientific discourse is evolving—despite our sense that it is fixed and highly regulated, it is not the same today as in the 1970s or 1940s or 1910s. One aspect of its current state is the growth of nonstandard forms of English, as researchers from an increasing array of non-anglophone and non-Western nations enter into the production of scientific writing. The next few decades will see these forms become more common and pressure mount for more journals to accept them, again within distinct limits. Periodicals in some non-anglophone nations that already publish papers in nonstandard English will also grow in stature over time and may even come to compete for status with established journals. Those publications that adapt and that draw contributions from a highly international community may well emerge as the most successful. Editors and editorial boards, therefore, might reevaluate their standards periodically and consider a degree of flexibility as an advantage.

This might seem a bit radical to some. In the first paper I worked on for Aziz, I left some of his own phrases—for instance, "These carbonates missed final understanding of their origin"; "Multiple porosities are observed but never yet explained"—out of a sense that the document was his and should retain something of his touch. These, of course, were all changed to "correct" usage ("The origin of these carbonates remains undetermined"; "Multiple porosity systems in these rocks are as yet unexplained"), with the comment from reviewers that "improper English" and "grammatical errors" needed to be addressed. None of this bothered Aziz in the least, yet for their part the reviewers clearly felt that they were defenders of higher standards, keepers of the flame. This is never wrong or self-aggrandizing. It is, however, too narrow for a world in which science is becoming fully global and therefore filled with diverse voices. As the evolution of Greek, Latin, and Arabic shows, norms are both essential and dynamic. Intelligibility must be maintained, but this does not argue for rigid standards.

A second point: Aziz did not want his articles published in his own country's journals, even if they were published in English. The *Egyptian Journal of Geology*, for example, one of the most prestigious earth science periodicals in the country, was never his first choice. To gain a research post at a high-level academic institution, it was far better that his work appear in a widely recognized international journal. Thus, there is a two-way problem between national and international science, with authors who focus on one or the other having no easy bridge between. High-quality science should be allowed, even counseled, to appear in both publishing realms, and in more than one language. It should be possible, that is, for researchers to publish their papers both in their native tongue for a national journal and in English for an international journal and audience. The reverse sequence should also be allowed and normalized. Publishing a translation of a paper should not be considered plagiarism or fraud (in the literary world, a translation is often considered a new text, even an original of its own). The problem of priority ownership of data, figures, and other nontextual material would have to be worked out. Current copyright and intellectual property law favors it remaining with the venue of first appearance. This should not, however, prevent permission being granted for its republication, with proper acknowledgment. The issue, at base, is not one of ownership or legacy investment but of science itself—the value of the information as communal capital.

Teaching and Learning English: The Core Issue

Above all, issues of linguistic inequity can be addressed through language training, as noted above. There is no substitute for this. Opening up international scientific discourse to a diversity of lingua francas isn't a workable or advisable option (even if it were possible, it would require many years to achieve, thus defeating the purpose). If English has become a needed skill for scientists who wish to participate at an international level, then it should be taught and learned as such, that is, fully integrated into the science curriculum. I have stated this a number of times in the present book. Here is the place to discuss what it means.

First and most important, it means unburdening English of its association with any specific country and treating it as a skill that anyone can learn. It means understanding that acquiring such a skill takes significant time, normally six to eight years, and thus requires investment, consistency, and patience. Second, English as a foreign language (EFL) in science needs to be considered a core subject, similar to mathematics or other fundamental parts of training. Any stigma associated with English or one's felt inability to learn it would be reduced were the language regarded as a normal part of science itself. More attention should therefore be paid to the quality and effectiveness of EFL teaching and everything that surrounds it—funding; teacher selection and training; instructional content; teaching methods; and student access, attitudes, and success. Most of the inequities in English ability begin here, after all; thus, anything that can improve this dimension of scientific training will be of great help. One type of improvement would likely come from abandoning the native-speaker model as the standard for student performance: accented speech should be considered normal, acceptable (as an indicator of origins, identity). Again, full intelligibility, especially to those from other countries, constitutes the only real, functional standard in a world with so many speakers from so many different linguistic backgrounds.

Beyond these basic ideas, what methods have proved effective in actual practice? What has worked best for the teaching of English to nonnative speakers? Global comparisons among countries, as well as individual experience, argue consistently that the highest success has been achieved in Scandinavia and the Netherlands. These nations have outperformed all others in teaching their students English, having done so since at least the early 1990s, without compromising their identities as Swedes, Danes, or Finns, and without sacrificing other subject areas, including other foreign languages. There are caveats to mention and discuss, however. These are all multilingual western European nations, with histories of contact with Indo-European tongues. The phonologies of Swedish, Norwegian, Icelandic, and Danish overlap with that of English, allowing EFL learners to sound more like native speakers than students from, say, France, India, or Japan. This is a distinct advantage, on one level. It brings status capital (so to speak) and betokens functional superiority. Yet in realistic terms, the ability to sound like an Englishman or an American makes little difference in terms of scientific performance; strong accents don't inevitably weaken the influence of one's work and reputation, as our story about Andre Geim in chapter 3 makes plain. Moreover, the overlap in phonology doesn't apply to Finnish, which is in a separate language family (Uralic, with affinities to Estonian, Hungarian, and several native tongues in Russia) from the Germanic group of tongues that include the other Scandinavian languages and Dutch.

It is said that children in these countries grow up watching and listening to popular media (TV, movies, music, computer games, and so on) from the United Kingdom and the United States; they are thus exposed to the sounds of English from an early age. But this is true in many parts of the world, where programs are not routinely dubbed. A more substantive advantage for these countries is that they have been multilingual for most or all of the modern era and largely view themselves in this way. It is part of the self of urban, educated Norwegians, Finns, Icelanders, Danes, Swedes, and Dutch to be "plurilingual" (the preferred term of the Council of Europe). This doesn't mean that every person in each nation speaks two or three languages, though they have studied them. But for people in the intellectual and professional classes, this is nearly always true. In every one of these countries, English is a compulsory subject starting in primary school. In Sweden and Denmark, German was widely studied before the arrival of English, particularly in the sciences, with French also fairly often taught. In other words, this part of Europe has long acquaintance with teaching, learning, and using foreign tongues.

Yet again, the question isn't about the advantages these nations may have, but the approaches they have devised for their success, and the degree to which their methods are transportable to other countries. To that end, here are some of the key elements in the approaches used in Scandinavia and the Netherlands:[4]

- A highly trained corps of teachers is viewed as absolutely essential. In none of these countries is the teacher required to be a native speaker or the native-speaker model held up as the final and only standard. Mother-tongue speakers of English are generally not imported to work in schools and universities. Rather, the primary emphasis is on well-trained, experienced teaching by natives of each country. Teaching ability, not native-like fluency, is the priority.
- Teaching in general, and the teaching of languages in particular, is highly valued as a profession. In Finland, for example, candidates chosen for university programs in teacher education are among the top high school students—admission is extremely competitive. Rigorous and demanding training is the rule. For teachers of English, study and work abroad in one or more anglophone countries are common. Quality of teaching is a distinct focus; it is understood that not everyone is suited to the profession, that teaching is closely tied to personality as well as learned techniques.
- Teachers are familiar with, and make use of, modern technology in the classroom to promote students' exposure to English. This includes the Internet, video, computers, and more. In countries like the Netherlands and Denmark, where English-language televi-

sion programs, films, computer games, and music are ubiquitous, teachers make use of children's familiarity with these media by integrating them into their lessons. In this way, early familiarity with English words and expressions acquired at home (again, from imported television programs, movies, and so on) is endorsed and treated as helpful experience.

- English is a core subject in the curriculum, from primary school to high school. This is consistent throughout each country. English courses are not reduced or eliminated during periods of economic recession or when governments, local and national, change party hands (as often happens with foreign languages in the United States). This type of treatment would make no sense in these countries. The success of a school or educational program, in fact, is evaluated on the basis of its English teaching (students' performance, for instance) as much as on other compulsory subjects, such as science, mathematics, and history.
- Because of these realities, English is placed in a wholly different category from other foreign languages and should not be compared with them. Rather than a subject of choice, it is treated as a required competence. National systems of assessment now include English, along with a few "marker" subjects, such as mathematics. In Norway, for example, national exams given in grades 4, 7, and 10 all test Norwegian, mathematics, and English. Similar tests are given in Iceland, Sweden, and Denmark. Moreover, the experience of learning English is used as a basis for introducing other foreign languages at later stages of a student's education.
- English-language study begins between the ages of six and eight. These classes are held every day of the school week, like other core subjects. Class hours for English are longer than those for social studies and cultural subjects, but shorter than those for the national language (in the Netherlands, for example, 400 hours/year, compared with 160 for "culture and the arts" and 480 for Dutch).[5] Teachers speak in English much of or nearly all the time in the primary grades (though not in the first year), then all the time by secondary school, with additional material such as grammar, discussion of readings, and question-and-answer sessions conducted entirely in English as well. Known as Content and

Language Integrated Learning (CLIL) and, in the United States, as "immersion," this approach to language learning has been strongly recommended by EFL theorists and applied linguists, and by policy-making organizations such as the European Union. It has been practiced in Scandinavia since the late 1970s and early '80s.

- English is also sometimes used in other, nonlanguage classes, particularly science, during the secondary grades. This somewhat spontaneous use of CLIL increases students' confidence in their own capability and also, through the teacher, models the use of English as a functional skill (rather than as cultural knowledge of inner-circle countries). It also prepares students for more routine CLIL at the university level. The approach overall has been shown to give learners more varied contact with the target language and to strengthen their progress toward multilingualism.
- A major part of English learning in the classroom is the focus on listening comprehension and oral performance. Conversation is stressed equally with reading and writing, and sometimes is given special attention. This emphasis is based on the observation that personal communication, the ability to actually perform as an English speaker, very often provides a student with the greatest sense of immediate pride and confidence. For this reason, teachers may allow imperfect grammar in conversation to help students avoid becoming self-conscious. Though successful, this does not prepare learners for actual professional work, particularly in technical fields, where specialized vocabulary is essential. Such preparation comes at the university level, from courses taught in English in the sciences and other fields.
- A recent trend aimed to accommodate the more highly motivated and performing students is full bilingual education. A growing number of primary and secondary schools are implementing such programs, with a complete curriculum in English. The number of such institutions is not large, and the chance this approach will become dominant in any sense seems slight, yet it fills a gap that existed for teaching higher-level students. Entrance to these programs is quite selective, making them elite in nature. These schools tend to have international connections and programs of study abroad (in the United Kingdom and United States, for example).

- Entrance to university programs in the sciences often requires the results from an international English-proficiency test, because most science courses are now taught in this language. This full English experience can prove challenging to first-year students as they adjust to the use of this language in a range of specialized subjects. Teachers may therefore purposely introduce technical terminology in both the native language and English at the beginning, then move to all English. By the third year, students are comfortable with English scientific discourse and are extending their specialized knowledge and vocabulary in their major field of study. When they enter graduate school, they are already or close to being fluent readers, speakers, and—very important—writers. Nearly all master's and doctoral theses are written in English, and students are strongly encouraged to write up their research for publication in an international journal (therefore in English).

Any or all of these approaches to English education could be adopted in other national settings, though in some cases this would require changes in the existing culture of education. None of these approaches, however, is singular to the culture, history, politics, or economic capability of northern Europe. Though they are part of an integrated educational system that has developed over time, they are focused on a single subject—English as a foreign language—and could serve as helpful advice for other countries, including emerging and poor ones. True, the reliance on up-to-date media in the classroom would not be possible in many instances. Yet the emphasis on quality of teaching, on starting English instruction early in the primary grades, on treating it as a core subject, and on incorporating experience with English in daily life into classroom learning are all approaches that depend on people, not technology. Ditto the use of the CLIL approach and the treatment of English less as a "foreign language" (a cultural carrier of American and Western society) than as a required skill for scientific work and publication.

Doubtless there will be obstacles. Attitudes toward English may need to change. Certain innovative adaptations, partly needed at an earlier phase in northern Europe (its period of debate over English and domain loss, discussed in chapter 4), may be needed along these lines. There are no guarantees, in other words—but there are strong probabilities. Not all the items in the list given above need be instituted at the same time. And except for teacher training, the most important aspects can be implemented without major expense or modification to the rest of the existing curriculum. A good teacher who uses CLIL can go a long way; consistent application of this technique from primary to secondary school would likely be enough to begin a major trend of improvement. Making English a widespread skill among professionals in most of the world's nations, particularly those wanting to further their science and technology capabilities, is hardly an impossible or overly ambitious goal.

No More Mono-Anglophonism?

Ideally, but also practically, the viewpoint that stresses English as a basic scientific skill has its flip side. As an ideal, every researcher might know how to employ several tongues, including English, as an expression of being a complete participant in an embracing, transnational, even planetary enterprise. This sounds nice, but is less than convincing. Practically, however, having access to scientific work and thought in only one language, while the rest of the world is able to utilize two or three, must be counted a disadvantage, even if that single tongue is global in extent. A native English speaker who can read material in another major language of science—Chinese, Spanish, Russian, Portuguese, German, French, or Japanese—is in a much superior position to a monolingual speaker.

Researchers today who have this capability often serve as critical mediators in multinational projects. They are more able to create or facilitate collaborations and to be leaders in related research, addressing problems that may arise not only in communication but in cultural misunderstandings and differing expectations. A global tongue will not erase all differences in culture. A physical chemist from Seoul will not abandon her sense of hierarchy or trained relationship to authority simply because she speaks English. The daily operations of research, from the role of the individual to the structure of organizations, reflect the society in which they occur. When it comes to the actual work of collaboration, therefore, mismatches of practice often happen. A global tongue can make these situations both better and worse, since it can disguise through seeming agreement a disconnect in expected behavior. This has often been observed in the case of East Asian researchers, whose cultural ways of expressing doubt, agreement, and criticism are often quite indirect and can be easily misinterpreted when translated directly into English. Scientists familiar with Chinese or Japanese culture are able to prevent these types of situations and help collaborative work proceed more smoothly.

In another sense, monolingual speakers of English, particularly those who have never studied another language to any great extent, can find certain important realities difficult to understand (or easy to ignore). They may be less tolerant of nonstandard forms of English or less able to comprehend speakers with strong accents. They may be unwilling to accept their own minority status in the global realm of English speakers and the inevitable varieties of their own language that legitimately exist. These problems create the potential for an appearance of insouciance, even arrogance—something likely to provoke animosity in some researchers from countries that suffered under colonialism. In this regard, it seems appropriate to consider the argument made by the Council of Europe in its *Guide for Language Education Policies*. Plurilingualism, it states,

> is not simply a functional necessity: it is also an essential component of democratic behavior. Recognition of the diversity of speakers' plurilingual repertoires should lead to linguistic tolerance and thus to respect for linguistic differences: respect for the linguistic rights of individuals and groups . . . respect for freedom of expression . . . respect for the diversity of languages for inter-regional and international communication. . . . Language teaching, the ideal locus for intercultural contact, is a sector in which education for democratic life in its intercultural dimensions can be included in education systems.[6]

We shouldn't think, however, that all this simply means constructing a better scientist-citizen of the global research republic. There is no substitute for intercultural skills. Science needs, and indeed has, a global tongue, but such a language cannot flatten the world. As long as national science is a vital part of the scientific enterprise, monolingualism is itself a form of isolation.

Final Words

Writing at the threshold of the Scientific Revolution, as one of its great founders, Galileo famously stated that the universe was written by God in the language of mathematics. He delivered these words, nevertheless, in his book *Il Saggiatore* (The Assayer; 1623), which he wrote in Italian, hoping it would attract a wider audience, at least on his home peninsula. Despite his skill in Latin, Galileo wielded the vernacular partly to signal the coming death of the language as a lingua franca. Having undergone no small trouble at the hands of the learned professoriate and priesthood, he employed his native Italian as a type of rebellion. Latin was to him less the language of science writ large than that of a small and inbred minority, whose conventional thinking was a hurdle to new ideas and discoveries. Thus began a new era. Latin soon failed, and science went in search of a new voice.

Three centuries later, Albert Einstein began his scientific career in German and ended it in English. His first writings appeared in print in 1901, more than a decade before World War I, and his final papers were published in 1955, the year of his death, ten years after World War II. During this period, the center of Western science moved across the Atlantic, from Germany to the United States, and began to find its new speech, which by the end of the century would be the world's voice, too. Einstein had a hard time with English at first (he was a notoriously bad speller). He spoke with a strong German accent. Yet he learned to write simple, sometimes elegant, English sentences. In school, he had studied French and Italian, which did not help him enormously after 1933, when in the wake of Hitler's rise to power he determined to live and work in America. Like Galileo, history made of him a multilingual harbinger of science's own linguistic future. Were both of these men alive today, and in the full of their careers, they would be able to talk with each other in a foreign but unifying tongue that has surpassed anything they might have imagined in their own time. English, as well as physics, would bring them together.

NOTES

Chapter 1

1. David Graddol, *English Next* (London: British Council, 2006); http://www.britishcouncil.org/learning-research-englishnext.htm (accessed July 2, 2012).
2. Jonathan Adams, Christopher King, and Nan Ma, *China: Research and Collaboration in the New Geography of Science*, Global Research Report, Thomson Reuters, November 2009; http://researchanalytics.thomsonreuters.com/grr/ (accessed June 22, 2012).
3. National Science Board, *Science and Engineering Indicators 2012*, chapter 2; http://www.nsf.gov/statistics/seind12/c2/c2h.htm.
4. A concise overview of publishing trends in Chinese science can be found in the article by David Cyranoski, "Strong Medicine for China's Journals," *Nature* 467 (2010): 261. Online September 15, 2010: http://www.nature.com/news/2010/100915/full/467261a.html.
5. Jeffrey Gil, "A Comparison of the Global Status of English and Chinese: Towards a New Global Language?," *English Today* 27, no. 1 (March 2011): 52–58.
6. Ibid., 53.
7. See for example Jerry Guo, "America's Chinese Problem," *Newsweek*, December 6, 2010; http://www.thedailybeast.com/newsweek/2010/12/06/not-much-progress-in-america-s-chinese-problem.html.
8. Joseph Lo Bianco, "English at Home in China: How Far Does the bond Extend?," in *China and English: Globalisation and the Dilemma of Identity*, ed. Joseph

Lo Bianco, Jane Orton, and Gao Yihong (Bristol: Multilingual Matters, 2009), 192–211; quotation is from p. 206.

9. Among the strong proponents of these ideas have been Tove Skutnabb-Kangas, Robert Phillipson, and Suresh Canagarajah. See for example Tove Skutnabb-Kangas, "Linguistic Genocide," in *Encyclopedia of Genocide and Crimes against Humanity*, ed. Dinah Shelton (New York: Macmillan, 2005), 653–54, and "'Do Not Cut My Tongue, Let Me Live and Die with My Language,' A Comment on English and Other Languages in Relation to Linguistic Human Rights," *Journal of Language, Identity and Education* 3, no. 2 (2004): 127–34; Robert Phillipson, *Linguistic Imperialism* (Oxford: Oxford University Press, 1992) and *Linguistic Imperialism Continued* (London: Routledge, 2010); and Suresh Canagarajah, *Resisting Linguistic Imperialism in English Teaching* (Oxford: Oxford University Press, 1999).

10. "Vision and Mission," King Abdullah University of Science and Technology; http://www.kaust.edu.sa/about/vision_mission.html#mission (accessed June 21, 2012).

11. Much of this history is covered in David Crystal, *English as a Global Language* (Cambridge: Cambridge University Press, 1997). A few additional aspects are treated, in meandering fashion, by Robert B. Kaplan, "English—the Accidental Language of Science?," in *The Dominance of English as a Language of Science: Effects on Other Languages*, ed. Ulrich Ammon (Berlin: Mouton de Gruyter, 2001), 3–26.

12. Seth Mydans, "The World: Vietnam Speaks English with an Eager Accent," *New York Times*, May 7, 1995; http://www.nytimes.com/1995/05/07/weekinreview/ the-world-vietnam-speaks-english-with-an-eager-accent.html?scp=1&sq=vietnam%20speaks%20english%20with%20an%20eager%20accent&st=cse.

13. "Disney's Schools in China: Middle Kingdom Meets Magic Kingdom," *Economist*, August 26, 2010; http://www.economist.com/node/16889262.

14. Studies that have tracked publication and citation in the humanities over the past three decades routinely show a growth in the overall percentage of articles produced in English, though they also confirm the continued importance of material in other languages as well. See for example Charlene Kellsey and Jennifer E. Knievel, "Global English in the Humanities? A Longitudinal Citation Study of Foreign-Language Use by Humanities Scholars," *College and Research Libraries* 65 (May 2004): 194–204.

15. The literature in this area is now vast. It is based partly on the rise of China as a possible global economic competitor with the United States and partly on the

perception of other rapidly developing nations lowering American preeminence, too. A few influential sources along these lines would include Fareed Zakaria, *The Post-American World* (New York: Norton, 2009); Niall Ferguson, *Colossus: The Rise and Fall of the American Empire* (New York: Penguin, 2005); Robert Kagan, *The Return of History and the End of Dreams* (New York: Vintage, 2009); and Eamonn Fingleton, *In the Jaws of the Dragon: America's Fate in the Coming Era of Chinese Hegemony* (New York: Thomas Dunne, 2008).

16. *EF EPI English Proficiency Index*, English First, 2011; http://www.ef.com/epi, p. 7.
17. On the phenomenon of world Englishes, the classic reference is Braj Kachru, ed., *The Other Tongue: English across Cultures*, 2nd ed. (Urbana: University of Illinois Press, 1992). A more recent treatment can be found in Andy Kirkpatrick, *World Englishes: Implications for International Communication and English Language Teaching* (London: Cambridge University Press, 2007). Interested readers should also consult the journal *World Englishes*.
18. See Scott L. Montgomery, *The Powers That Be: Global Energy for the Twenty-First Century and Beyond* (Chicago: University of Chicago Press, 2010), especially chapter 13, "Geopolitics and Energy."
19. Song Jian, "Awakening: Evolution of China's Science and Technology Policies," *Technology in Society* 30, nos. 3–4 (2008): 235–41. See also the 2008 interview with Chinese premier Wen Jiabao, himself a geologist: Hao Xin and Richard Stone, "China's Scientist Premier," *Science*, October 2008, 362–64.
20. *Science and Engineering Indicators 2012*, chapter 4. See also J. Thomas Ratchford and William A. Blanpied, "Paths to the Future for Science and Technology in China, India and the United States," *Technology in Society* 30, nos. 3–4 (2008): 211–33.
21. *World Data on Education: VII Edition 2010/11—Brazil*, United Nations Educational, Scientific and Cultural Organization (UNESCO), April 2010, p. 5; http://www.ibe.unesco.org/fileadmin/user_upload/Publications/WDE/2010/LATIN_AMERICA_and_the_CARIBBEAN/Brazil/Brazil.pdf (accessed July 2, 2012).
22. Richard C. Levin, "Top of the Class: The Rise of Asia's Universities," *Foreign Affairs* (May–June 2010): 63–75.
23. See for example Philip G. Altbach, Liz Reisberg, and Laura E. Rumbley, *Trends in Global Higher Education: Tracking an Academic Revolution*, United Nations Education, Scientific, and Cultural Organization (UNESCO), 2009, http://unesdoc.unesco.org/images/0018/001831/183168e.pdf (accessed July 1, 2012); and Douglas Gilman, *The New Geography of Global Innovation*, Global Markets Institute,

Goldman Sachs, September 20, 2010, http://www.innovationmanagement.se/wp-content/uploads/2010/10/The-new-geography-of-global-innovation.pdf (accessed June 23, 2012).

24. Ben Wildavsky, *The Great Brain Race: How Global Universities Are Reshaping the World* (Princeton: Princeton University Press, 2010).

25. For a reasonable sampling of this literature, see the following books: Ulrich Ammon, ed., *The Dominance of English as a Language of Science: Effects on Other Languages* (Berlin: Walter de Gruyter, 2001); Christian Mair, ed., *The Politics of English as a World Language* (Amsterdam: Editions Rodopi B.V.); and Suresh Canagarajah, *A Geopolitics of Academic Writing* (Pittsburgh: University of Pittsburgh Press, 2002). Thoughtful, in-depth reviews of related issues are offered by Ulrich Ammon, "Language Planning for International Scientific Communication: An Overview of Questions and Potential Solutions," *Current Issues in Language Planning* 7, no. 1 (April 2006): 1–30, and Rainer Enrique Hamel, "The Dominance of English in the International Scientific Periodical Literature and the Future of Language Use in Science," *AILA Review* (Association Internationale de Linguistique Appliquée) 20 (2007): 53–71, plus references therein. For a briefer discussion, see Bonnie Lee La Madeleine, "Lost in Translation," *Nature* 445, no. 7126 (2007): 454–57.

Chapter 2

1. Matthew Parris, "The Global Spread of English Is a Seismic Event in Human History," *Times* (London), January 5, 2005; http://www.timesonline.co.uk/tol/comment/columnists/matthew_parris/article412560.ece.

2. David Crystal, *English as a Global Language* (Cambridge: Cambridge University Press, 1997), ix; and *English as a Global Language*, 2nd ed. (London: Cambridge, 2003), x.

3. David Graddol, *The Future of English* (London: British Council, 1997), and Graddol, *English Next* (London: British Council, 2006).

4. *Ethnologue: Languages of the World* is a reference work published by SIL International (Summer Institute of Linguistics), a nonprofit, Christian organization headquartered in Dallas, Texas. Founded in 1934, SIL provides linguistic services to nations and groups worldwide. It does not engage in proselytizing activities directly, but provides language information for missionaries. *Ethnologue* remains unparalleled in its global coverage and is updated every four years (see http://www.ethnologue.org/ [accessed July 2, 2012]).

5. See international census information provided by the United Nations Statistics Division, on its website for demographic and social data: http://unstats.un.org/unsd/demographic/default.htm (accessed July 2, 2012).
6. David Graddol, *English Next India* (London: British Council, 2010), 50, 68; http://www.britishcouncil.org/learning-english-next-india-2010-book.htm (accessed June 12, 2012).
7. See David Crystal, *Cambridge Encyclopedia of the English Language* (Cambridge: Cambridge University Press, 1995); Crystal, *English as a Global Language*; Braj B. Kachru, *Asian Englishes: Beyond the Canon* (Seattle: University of Washington Press, 2005); Graddol, *English Next*. For estimates on Asia, see Kingsley Bolton, "English in Asia, Asian Englishes, and the Issue of Proficiency," *English Today* 94, no. 3 (2008): 3–12.
8. Yun-Kyung Cha and Seung-Hwan Ham, "The Impact of English on the School Curriculum," in *The Handbook of Educational Linguistics*, ed. Bernard Spolsky and Francis M. Hult (Oxford: Blackwell, 2008), 313–27; quotation is from p. 313.
9. Joseph Nye, *Bound to Lead: The Changing Nature of American Power* (New York: Basic Books, 1991) and, more recently, *Soft Power: The Means to Success in World Politics* (New York: Public Affairs Press, 2005).
10. Cha and Ham, "Impact of English on School Curriculum." See also by these authors "Educating Supranational Citizens: The Incorporation of English Language Education into Curriculum Policies," *American Journal of Education* 117, no. 2 (February 2011): 183–209.
11. Institute of International Education, "Project Atlas, Atlas of Student Mobility," http://www.iie.org/en/research-and-publications/project-atlas; OECD Directorate for Education, "Education at a Glance 2011," http://www.oecd.org/document/2/0,3746,en_2649_39263238_48634114_1_1_1_1,00.html; Council of Graduate Schools, "International Graduate Admissions Survey," http://www.cgsnet.org/Default.aspx?tabid=172; and Nathan E. Bell, "Findings from the 2012 CGS International Graduate Admissions Survey, Phase I: Applicaitons," http://www.cgsnet.org.
12. Websites in English for these universities are as follows: Nebrija Universidad, http://www.nebrija.com/master-mba/university/index.html, and University of Helsinki, http://www.helsinki.fi/university/index.html.
13. Tohoku University, "Inoue Plan 2007—Our Path toward Becoming a World Class University—Revised for 2009"; http://www.bureau.tohoku.ac.jp/president/open/plan/Inoue_Plan_2009.pdf (accessed June 22, 2012).
14. See for example "Chinese Universities to Use Textbooks Written in English,"

Xinhua, October 22, 2001; http://news.xinhuanet.com/english/20011022/465118.htm, and "English-Taught Programs at Chinese Universities," China.org.cn, October 18, 2006; http://www.china.org.cn/english/LivinginChina/184768.htm.

15. David Graddol, "Global English," Open2.net, August 7, 2005, Open University, Manchester, UK; http://www.open2.net/healtheducation/education/globalenglish.html.
16. "English Is Coming," *Economist*, February 12, 2009; http://www.economist.com/node/13103967.
17. For discussion of the new linguistic policy, see Stephanie McCrummen, "Rwandans Say Adieu to Français," *Washington Post*, October 28, 2008; http://www.washingtonpost.com/wp-dyn/content/article/2008/10/27/AR2008102703165.html, and Chris McGreal, "Rwanda to Switch from French to English in Schools," *Guardian* (Manchester), October 14, 2008; http://www.guardian.co.uk/world/2008/oct/14/rwanda-france.
18. See Roméo Dallaire and Samantha Power, *Shake Hands with the Devil: The Failure of Humanity in Rwanda* (Cambridge, MA: Da Capo Press, 2004).
19. Pete Browne, "Tapping Methane at Lake Kivu in Africa," *New York Times*, September 23, 2011; http://green.blogs.nytimes.com/2010/01/26/tapping-methane-at-africas-lake-kivu/. Information about, and access to, the report by Eawag Aquatic Research is available at http://www.eawag.ch/forschung/surf/gruppen/kivu/methane_harvesting/index_EN (accessed July 7, 2012).
20. Papers and summary reports from this symposium are available from the Digital Imaging and Remote Sensing Laboratory at the Rochester Institute of Technology, http://dirs.cis.rit.edu/node/270 (accessed July 2, 2012).
21. Francisco Moreno Fernandez and Jaime Otero Roth, "Demografia de la lengua español," Instituto Complutense de Estudios Internacionales, 2006; http://eprints.ucm.es/8936/1/DT03-06.pdf (accessed July 7, 2012).
22. Figures quoted in this paragraph are from "Internet World Stats: Usage and Population Statistics," Miniwatts Marketing Group; http://www.internetworldstats.com/stats7.htm (accessed June 10, 2012). See also Peter Gerrand, "Estimating Linguistic Diversity on the Internet: A Taxonomy to Avoid Pitfalls and Paradoxes," *Journal of Computer-Mediated Communication* 12, no. 4 (2007), article 8; http://jcmc.indiana.edu/v0112/issue4/ gerrand.html. Regarding statistics relevant to the digital divide, see International Telecommunications Union, "ICT Statistics"; http://www.itu.int/ITU-D/ict/statistics/ (accessed January 10, 2012).

23. Internet World Stats, http://www.internetworldstats.com/stats7.htm (accessed June 26, 2012).
24. Hossein Bidgoli, ed., *The Internet Encyclopedia* (New York: Wiley, 2003), 1:434.
25. Internet World Stats, http://www.internetworldstats.com/stats7.htm (accessed June 26, 2012).
26. International Energy Agency, *World Energy Outlook 2011* (Paris: International Energy Agency, 2011).
27. World Bank, *World Development Indicators*, World Bank Group, New York, 2011; http://data.worldbank.org/data-catalog/world-development-indicators.
28. See Marcos Aguiar, Vladislav Boutenko, David Michael, Vaishali Rastogi, Arvind Subramanian, and Yvonne Zhou, *The Internet's New Billion: Digital Consumers in Brazil, Russia, India, China, and Indonesia*, Boston Consulting Group, 2010; http://www.bcg.com/expertise_impact/Capabilities/Globalization/PublicationDetails.aspx?id=tcm:12-58652 (accessed June 28, 2012).
29. "Top UN Official Stresses Need for Internet Multilingualism to Bridge Digital Divide," UN News Centre, December 14, 2009; http://www.un.org/apps/news/story.asp?NewsID=33256&Cr=akasaka&Cr1=.
30. "Statistical Summaries," *Ethnologue: Languages of the World*, 16th ed. (Dallas, TX: Summer Institute of Linguistics International, 2009); http://www.ethnologue.com/ethno_docs/distribution.asp?by=size. See also David Crystal, *Language Death* (London: Cambridge University Press, 2000), 19–26.
31. These numbers are taken from Anthony C. Woodbury, "What Is an Endangered Language?," Linguistic Society of America; http://www.lsadc.org/info/ling-faqs-endanger.cfm (accessed July 1, 2012). Similar figures exist in a number of other sources; see for example the website Native Languages of the Americas: List of Native American Indian Tribes and Languages, which provides information on over 190 individual tongues, at http://www.native-languages.org/languages.htm#alpha (accessed July 1, 2012).
32. See for example Lenore A. Grenoble and Lindsay J. Whaley, *Endangered Languages: Current Issues and Future Prospects* (Cambridge: Cambridge University Press, 2010), and *UNESCO Interactive Atlas of the World's Languages in Danger*; http://www.unesco.org/culture/ich/index.php?pg=00206 (accessed June 7, 2012).
33. There are now several projects directly involved in archiving endangered languages, with the specific goal of their protection. These include such efforts as Living Tongues Institute (http://www.livingtongues.org/), the Rosetta Project (http://rosettaproject.org/), and Enduring Voices (National Geographic Society,

http://travel.nationalgeographic.com/travel/enduring-voices/), as well as a growing number of projects focused on revitalizing specific languages, such as Native Languages of the Americas: Endangered Language Revitalization and Revival endeavor (http://www.native-languages.org/revive.htm) and American Indian Language Development Institute (http://aildi.arizona.edu/).

34. See for example Norimitsu Onishi, "With Casino Revenues, Tribes Push to Preserve Languages and Cultures," *New York Times*, June 16, 2012; http://www.nytimes.com/2012/06/17/us/chukchansi-tribe-in-california-pushes-to-preserve-language.html?pagewanted=1&_r=1&nl=todaysheadlines&emc=edit_th_20120617.
35. Several widely cited sources in this vein are Tove Skutnabb-Kangas, *Linguistic Genocide in Education* (London: Routledge, 2000); Robert Phillipson, *Linguistic Imperialism* (London: Oxford University Press, 1992); and Suresh Canagarajah, *Resisting Linguistic Imperialism in English Teaching* (London: Oxford University Press, 1999).
36. Salikoko Mufwene, "Colonisation, Globalisation, and the Future of Languages in the Twenty-First Century," *International Journal on Multicultural Societies* 4, no. 2 (2002): 162–93.
37. For an interesting discussion on this topic and related issues, see the interview with David Crystal by Katarina Basulic, "A Nation without a Language Is a Nation without a Heart," *Belgrade English Language and Literature Studies* 1 (2009): 231–54. The article is available on David Crystal's website: http://www.davidcrystal.com/David_Crystal/articles.htm (accessed July 1, 2012).
38. Paul Bruthiaux, "Dimensions of Globalization and Applied Linguistics," in *Language as Commodity: Global Structures, Local Marketplaces*, ed. Peter K.W. Tan and Rani Rubdy (London: Continuum, 2008), 16–30.
39. Sam Roberts, "Listening to (and Saving) the World's Languages," *New York Times*, April 28, 2010; http://www.nytimes.com/2010/04/29/nyregion/2910st.html?pagewanted=all.
40. Nkonko M. Kamwangamalu, "When 2 + 9 = 1: English and the Politics of Language Planning in a Multilingual Society," in *The Politics of English as a World Language*, ed. Christian Mair (Amsterdam: Editions Rodopi, 2003), 235–46. More detailed information is available in Kamwangamalu's book *One Language, Multi-Layered Identities: English in a Society in Transition, South Africa* (New York: Blackwell, 2007).
41. Martin Moir and Lynn Zastoupil, eds., *The Great Indian Education Debate: Documents Relating to the Orientalist-Anglicist Controversy, 1781–1843* (Richmond, UK: Curzon Press, 1999).

42. For the full text of the Macaulay Minute and commentary, see M. S. Thirumalai, "Lord Macaulay, the Man Who Started It All, and His Minute," *Language in India* 3, no. 4 (April 2003); http://www.languageinindia.com/april2003/macaulay.html. A good brief discussion of the minute and its impact can also be found in Janina Brutt-Griffler, *World English: A Study of Its Development* (Bristol, UK: Multilingual Matters, 2002).
43. Bruce T. McCully, *English Education and the Origins of Indian Nationalism* (Gloucester, MA: Peter Smith, 1966); Steven Evans, "Macaulay's Minute Revisited: Colonial Language Policy in Nineteenth-Century India," *Journal of Multilingual and Multicultural Development* 23, no. 4 (2002): 260–81.
44. Makere University, January 2011 Speeches, 31st January–1st February, AET2011; http://mak.ac.ug/index.php?option=com_content&task=view&id=395&Itemid=219 (accessed June 12, 2012).
45. Department of Agriculture and Cooperation, *Annual Report 2010–2011* (Ministry of Agriculture, Government of India, March 2011), p. 3; http://agricoop.nic.in/docs.htm (accessed June 12, 2012).
46. Braj Kachru, *The Other Tongue: English across Cultures* (Urbana: University of Illinois Press, 1992).
47. A listing of objections to the Kachru model can be found in Jennifer Jenkins, *World Englishes: A Resource Book for Students* (London: Routledge, 2003), 17–18. A recent attempt to take many of these criticisms into account and produce a revision of the model can be found in Yasukata Yano, "The Future of English: Beyond the Kachruvian Three Circle Model," in *Global Englishes in Asian Contexts: Current and Future Debates*, ed. Kumiko Murata and Jennifer Jenkins (New York: Palgrave Macmillan, 2009), 208–25. The result, however, is a 3-D model that requires a good deal of explanation and thus seems unwieldy and not particularly clarifying.
48. See for example Sandra Lee McKay, *Teaching English as an International Language: Rethinking Goals and Perspectives* (New York: Oxford University Press, 2002).
49. Andy Kirkpatrick, *World Englishes: Implications for International Communication and English Language Teaching* (London: Cambridge University Press, 2007), 188.

Chapter 3

1. For biographical information about Andre Geim, see "The Nobel Prize in Physics 2010," Nobelprize.org, http://www.nobelprize.org/nobel_prizes/physics/

laureates/2010/geim.html, and the interview by Gali Weinreb, "Nobel Laureate: Life Sciences Suited to Small Countries," *Jerusalem Post,* November 20, 2010; http://www.jpost.com/LandedPages/PrintArticle.aspx?id=196080. The quotation about the brain drain is from the interview with Gali Weinreb.

2. David A. Kronick, *History of Scientific and Technical Periodicals: The Origin and Development of the Scientific and Technical Press, 1665–1790* (Lanham, MD: Rowman & Littlefield, 1976).
3. Ibid., chapter 3; James McClellan, "The Scientific Press in Transition: Rozier's Journal and the Scientific Societies in the 1770s," *Annals of Science* 36, no. 5 (September 1979): 425–49.
4. Quoted in Kronick, *History of Scientific and Technical Periodicals,* 100–101.
5. See for example Ulrich Ammon, *Ist Deutsch noch internationale Wissenschaftssprache? Englisch auch für die Lehre an den deutschsprachigen Hochschulen* [Is German still an international language of science? English even for teaching at German universities] (Berlin: de Gruyter, 1998); Michael Mabe and Mayur Amin, "Growth Dynamics of Scholarly and Scientific Journals," *Scientometrics* 51, no. 1 (2001): 147–62; Michael Mabe, "The Growth and Number of Journals," *Serials* 16, no. 2 (2003): 191–97; Robert May, "The Scientific Wealth of Nations," *Science,* February 7, 1997, pp. 793–96; and Peder Olesen Larsen and Markus von Ins, "The Rate of Growth in Scientific Publication and the Decline in Coverage Provided by Science Citation Index," *Scientometrics* 84, no. 3 (September 2010): 575–603. The first such study of wide influence, now a classic in the field, is Derek de Solla Price's *Science since Babylon* (New Haven, CT: Yale University Press, 1961).
6. Roswitha Reinbothe, *Deutsch als internationale Wissenschaftssprache und der Boykott nach dem Ersten Weltkrieg* (Frankfurt: Lang, 2006).
7. National Science Board, *Science and Engineering Indicators 2012,* p. 4–48, http://www.nsf.gov/statistics/seind12/c2/c2h.htm; and United Nations Education, Science, and Cultural Organization (UNESCO), *UNESCO Science Report 2010* (Paris: UNESCO Publishing, 2010), page 2, table 1; http://www.unesco.org/new/en/natural-sciences/science-technology/prospective-studies/unesco-science-report/.
8. *Science and Engineering Indicators 2012,* p. 4–43.
9. Percentages are calculated from data presented in *UNESCO Science Report 2010,* p. 8, table 2.
10. Quoted in "Andre Geim: In Praise of Graphene," *Nature,* published online October 7, 2010; http://www.nature.com/news/2010/101007/full/news.2010.525.html (accessed August 14, 2011).

11. *UNESCO Science Report 2010*, 5.
12. See Yaw Nyarko, "The Returns to the Brain Drain and Brain Circulation in Sub-Saharan Africa: Some Computations Using Data from Ghana," National Bureau of Economic Research Working Paper 16813, February 2011; http://www.nber.org/papers/w16813.
13. National Science Board, *Science and Engineering Indicators: 2010*, National Science Foundation, Arlington, VA, p. 5-32, fig. 5-20; http://www.nsf.gov/statistics/seind10/.
14. Chen Jia, "Nation Aims to Increase Talent Pool," *China Daily*, June 7, 2010; http://www.chinadaily.com.cn/china/2010-06/07/content_9940774.htm.
15. Caroline S. Wagner, *The New Invisible College* (Washington, DC: Brookings Institution Press, 2008).
16. Philippe Moguérou, "The Brain Drain of PhDs from Europe to the United States: What We Know and What We Would Like to Know," EUI Working Paper RSCAS No. 2006/11 (San Domenico di Fiesole, Italy: Robert Schuman Centre for Advanced Studies, European University Institute, 2006), 3.
17. Karen M. Dente, "Scientists on the Move," *Cell* 129 (April 6, 2007): 15-17.
18. National Science Board, *Science and Engineering Indicators: 2010*, p. 6-5.
19. Mariana Zanatta and Sergio Queiroz, "The Role of National Policies on the Attraction and Promotion of Multinational Enterprises' R&D Activities in Developing Countries," *International Review of Applied Economics* 21, no. 3 (2007): 419-435. Similar points are made in the report by the National Science Board, "Globalization of Science and Engineering Research" National Science Foundation, Arlington, VA, 2010); http://www.nsf.gov/statistics/nsb1003/.
20. Thomson Reuters, "The Thomson Reuters Journal Selection Process," Thomson Reuters, Science; http://thomsonreuters.com/products_services/science/free/essays/journal_selection_process/ (accessed June 20, 2012).
21. Ana María Cetto, José Octavio Alonso-Gamboa, and Saray Córdoba González, "Ibero-American Systems for the Dissemination of Scholarly Journals: A Contribution to Public Knowledge Worldwide," *Scholarly and Research Communication* 1, no. 1 (2010): 1-16.
22. Rainer Enrique Hamel, "The Dominance of English in the International Scientific Periodical Literature and the Future of Language Use in Science," *AILA Review* 20 (2007): 53-71.
23. Abhaya V. Kulkarni, Brittany Aziz, Iffat Shams, and Jason W. Busse, "Comparisons of Citations in Web of Science, Scopus, and Google Scholar for Articles Published

in General Medical Journals," *Journal of the American Medical Association* 302, no. 10 (2009): 1092–96.

24. For discussion of these trends and related criticism, see *Bibliometric Analysis in Science and Research: Applications, Benefits, and Limitations*, Conference Proceedings, Schriften des Forschungszentrums Jülich Reihe Bibliothek, vol. 11 (Jülich, Germany: Reihe Bibliothek, 2004).
25. *UNESCO Science Report* 2010, p. 10, table 3.
26. Ibid.; National Science Board, *Science and Engineering Indicators: 2010*, p. 5–31, table 5–14; Terry Commins, Warinthorn Songkasiri, Suvit Tia, and Bundit Tipakorn, "Science and Technology Research in Thailand: Some Comparisons from the Data regarding Thailand's Position in the Region Based on Volume of Published Work," *Maejo International Journal of Science and Technology* 2, no. 3 (2008): 508–15. For data on Ibero-American and Caribbean countries between 1973 and 2010, see G. A. Lemarchand, "The Long-Term Dynamics of Co-Authorship Scientific Networks, Iberoamerican Countries (1973–2010)," *Research Policy* 41, no. 2 (2012): 291–305.
27. The Royal Society, *Knowledge, Networks and Nations: Global Scientific Collaboration in the 21st Century*, Royal Society Policy Document 03/11 (London: Royal Society, 2011), 16–17. A similar pattern occurs for Web of Science data, which indicates more than a doubling in the proportion of collaborative publications between 1988 and 2003. See National Science Board, *Science and Engineering Indicators: 2006*, National Science Foundation, Arlington, VA; http://www.nsf.gov/statistics/seind06/?org=NSF, chapter 5. According to this issue of *Science and Engineering Indicators*, the number of nations with which US scientists collaborated rose to 173 (2003 figure), the highest of all countries, and "the average number of foreign addresses on U.S. scientific articles more than tripled" (p. 5–42).
28. G. Aad et al., "Charged-Particle Multiplicities in *pp* Interactions at $\sqrt{s} = 900$ GeV Measured with the ATLAS Detector at the LHC," *Physics Letters B* 688 (2010): 21–42.
29. Royal Society, *Knowledge, Networks and Nations*, p. 48, fig. 2.2.
30. Fraunhofer ISI Idea Consult, *The Impact of Collaboration on Europe's Scientific and Technological Performance*, Fraunhofer Institute Systems and Innovation Research, March 2009; http://ec.europa.eu/invest-in-research/ pdf/download_en/final_report_spa2.pdf.
31. See for example Caroline S. Wagner and Loet Leydesdorff, "Network Structure, Self-Organization, and the Growth of International Collaboration in Science," *Research Policy* 34, no. 10 (2005): 1608–18; José J. Ramasco, S. N. Dorogovtsev,

and R. Pastor-Santorras, "Self-Organization of Collaboration Networks," *Physical Review E* 70 (2004): 1–10; and Lemarchand, "The Long-Term Dynamics of Co-Authorship."

32. Relevant information can be found in National Science Board, *Science and Engineering Indicators: 2006*, p. 5–43.
33. The Royal Society, *Knowledge, Networks and Nations*, 49–51.
34. See for example Politimi E. Valkimadi, Drosos E. Karageorgopoulos, Harissios Vliagoftis, and Matthew E. Falagas, "Increasing Dominance of English in Publications Archived by PubMed," *Scientometrics* 81, no. 1 (2009): 219–23; Mohammad H. Biglu and Walther Umstätter, "The Editorial Policy of Languages Is Being Changed in Medline," *Acimed* 16, no. 3 (2007), http://bvs.sld.cu/revistas/aci/v0116_3_07/aci06907.htm; Michele Bedard, Jennifer L. Greif, and Todd Buckley, "International Publication Trends in the Traumatic Stress Literature," *Journal of Traumatic Stress* 17, no. 2 (2004): 97–101; Alvar Loria and Pedro Arroyo, "Language and Country Preponderance Trends in Medline and Its causes," *Journal of the Medical Library Association* 93, no. 3 (2005): 381–85; Mauricio L. Barreto, "Growth and Trends in Scientific Production in Epidemiology in Brazil," *Revista de Saúde Pública* 40 (2006): 79–85; S. Yamazaki and H. Zhang, "Internationalization of the English-Language Journals in Japan in Life Sciences," *Nippon Seirigaku Zasshi* 59, no. 2 (1997): 98–104; Matthew E. Falagas, Eufemia Fabritsi, Fotini C. Chelvatzoglou, and Konstantinos Rellos, "Penetration of the English Language in Science: The Case of a German National Interdisciplinary Critical Care Conference," *Critical Care* 9, no. 6 (2005): 655–56; and Andreas Dinkel, Hendrik Berth, Ada Borkenhagen, and Elmar Brahler, "On Raising the International Dissemination of German Research: Does Changing Publication Language to English Attract Foreign Authors to Publish in a German Basic Psychology Research Journal?," *Experimental Psychology* 51, no. 4 (2004): 319–28.
35. The data for this graph have been adapted from Tsunoda Minoru, "Les langues internationales dans les publications scientifiques et techniques," *Sophia Linguistica* 13 (1983): 144–55; Ammon, *Ist Deutsch noch Internationale Wissenschaftssprache?*; and National Science Board, *Science and Engineering Indicators 2012*.
36. Yun-Kyung Cha and Seung-Hwan Ham, "Educating Supranational Citizens: The Incorporation of English Language Education into Curriculum Policies," *American Journal of Education* 117, no. 2 (February 2011): 183–209.
37. Heinrich Zankl and Wolf-Christian Dullo, "Editorial Address of the Editor-in-Chief and the Editor-in-Chief Elect," *Geologische Rundschau* 82 (1993): 1–2.
38. Wolf-Christian Dullo, editorial, *Geologische Rundschau* 83 (1994): 1–2.

39. For a preliminary discussion of this phenomenon, see Scott L. Montgomery, *Science in Translation: Movements of Knowledge through Cultures and Time* (Chicago: University of Chicago Press, 2002), 57–62.

40. Ayse Gündogdu, F. Burcu Harmantepe, Zafer Karsli, and Gaye Dogan, "Elimination of Copper in Tissues and Organs of Rainbow Trout (*Oncorhynchus mykiss*, Walbaum, 1792) following Dietary Exposure," *Italian Journal of Animal Science* 10, no. 1 (2011): 1–5; quotation is from p. 1.

41. Qin Zou, Shujing Gao, and Qi Zhong, "Pulse Vaccination Strategy in an Epidemic Model with Time Delays and Nonlinear Incidence," *Advanced Studies in Biology* 1, no. 7 (2009): 307–21; quotation is from p. 307.

42. A. K. Abdel-Fattah, K. Y. Kim, and M. S. Fnais, "Slip Distribution Model of Two Small-Sized Inland Earthquakes and Its Tectonic Implication in North-Eastern Desert of Egypt," *Journal of African Earth Sciences* 61 (2011): 296–307; quotation is from p. 296.

43. Alan G. Gross, Joseph E. Harmon, and Michael S. Reidy, *Communicating Science: The Scientific Article from the 17th Century to the Present* (Anderson, SC: Parlor Press, 2009).

44. An example of such variation is the tendency for e-journals in the Philippines to include a brief objectives section immediately following the introduction. The remaining portions of these papers, however, follow the standard materials/methods, results, and discussion/conclusions structure.

45. Jonathan Adams and David Pendlebury, "Global Research Report: United States," Thomson Reuters, November 2010, p. 1; http://researchanalytics.thomsonreuters.com/grr/.

46. David Cyranoski, "Strong Medicine for China's Journals," *Nature* 467, no. 261 (September 15, 2010); http://www.nature.com/news/2010/100915/full/467261a.html (accessed July 7, 2012).

47. Vannevar Bush, "Science—The Endless Frontier; A Report to the President by Vannevar Bush"; http://www.nsf.gov.od/1/nsf50/vbush1945.htm.

48. Andre Geim, "Autobiography," Nobelprize.org (official website of the Nobel Prize); http://www.nobelprize.org/nobel_prizes/physics/laureates/2010/geim.html (accessed June 24, 2012).

Chapter 4

1. Scott L. Montgomery, *Science in Translation: Movements of Knowledge through Cultures and Time* (Chicago: University of Chicago Press, 2002).

2. A significant literature has come to surround this idea of language inequality or, in some writings, "injustice." One branch of this literature has tended to take a highly critical position toward English as a global tongue for science and advocates major changes in language planning and policy for the whole of international scientific publishing. While these writings are mostly by nonscientists and do not much consider, in any consistent manner, the opinions of researchers themselves, they nonetheless raise a number of issues that can be neither avoided nor ignored in any balanced view of the larger situation. A selection would include Ulrich Ammon, "Global English and the Non-Native Speaker: Overcoming Disadvantage," in *Language in the 21st Century*, ed. Humphrey Tonkin and Timothy Reagan (Amsterdam: Benjamins, 2003), 23–34; Ulrich Ammon, "Language Planning for International Scientific Communication: An Overview of Questions and Potential Solutions," *Current Issues in Language Planning* 7 (2006): 1–31; A. Suresh Canagarajah, *A Geopolitics of Academic Writing* (Pittsburgh: University of Pittsburgh Press, 2002); Bonnie Lee La Madeleine, "Lost in Translation," *Nature* 445 (2007): 454–55; R. E. Hamel, "The Dominance of English in the International Scientific Periodical Literature and the Future of Language Use in Science," *AILA Review* 20 (2008): 53–71; Humphrey Tonkin, "Language and the Ingenuity Gap," *Scientist* 22, no. 4 (2008): 1–10; Erin Bidlake, "Whose Voice Gets Read? English as the International Language of Scientific Publication," *E-pisteme* 1, no. 1 (2008): 3–21; Miguel Clavero, "'Awkward wording. Rephrase': Linguistic Injustice in Ecological Journals," *Trends in Ecology and Evolution* 25, no. 10 (2010): 552; and Charles Durand, *La Mise en Place des Monopoles du Savoir* (Paris: L'Harmattan, 2001). A second branch of the relevant literature focuses less on issues of policy and intervention and more on the attitudes of researchers, as well as giving a more fine-grained analysis of nonnative speakers' experiences in working with English. Some representative publications here include Gibson Ferguson, Carmen Pérez-Llantada, and Ramón Plo, "English as an International Language of Scientific Publication: A Study of Attitudes," *World Englishes* 30, no. 1 (2011): 41–59; Diane Belcher, "Seeking Acceptance in an English-Only Research World," *Journal of Second Language Writing* 16 (2007): 1–22; Mary Curry and Theresa Lillis, "Multilingual Scholars and the Imperative to Publish in English: Negotiating Interests, Demands and Rewards," *TESOL Quarterly* 38, no. 4 (2007): 663–88; John Flowerdew, "The Non-Anglophone Scholar at the Periphery of Scientific Communication," *AILA Review* 20 (2008): 14–27; Laura Landa, "Academic Language Barriers and Language Freedom," *Current Issues in Language Planning* 7 (2006): 61–81; and Ragnhild Ljosland, "English in Norwegian Academia: A Step toward Diglossia," *World Englishes* 26, no. 4 (2007): 395–410.

Finally, there are also discussions that attempt to look at linguistic inequity in terms of differences in English skills between developed and developing countries. An oft-cited example here is Françoise Salager-Meyer, "Scientific Publishing in Developing Countries: Challenges for the Future," *Journal of English for Academic Purposes* 7, no. 2 (2008): 121–32.

3. Eugene Garfield and Alfred Welljams-Dorok, "Language Use in International Research: A Citation Analysis," *Annals of the American Academy of Political and Social Science* 511 (1990): 10–24; quotation is from p. 10.
4. David Cyranoski, "Bird Flu Data Languish in Chinese Journals," *Nature* 430, no. 955 (August 26, 2004); http://www.nature.com/nature/journal/v430/n7003/full/430955a.html (accessed January 12, 2012).
5. Manuel R. Guariguata, Douglas Sheil, and Daniel Murdiyarso, "'Linguistic Injustice' Is Not Black and White," *Trends in Ecology and Evolution* 26, no. 2 (November 2010): 58–59, and Michael Hewings, "English Language Standards in Academic Articles: Attitudes of Peer Reviewers," *Revista Canaria de Estudios Ingleses* 53 (2006): 47–62.
6. See the summary of various ideas along these lines in Ammon (2007) and Salager-Meyer (2008). A strong voice for the plurilingual approach is Hamel (2007).
7. Aside from the evidence of my informal interviews with nonnative speakers as well as experience as editor, coauthor, and sometime scientific ghostwriter, these points are confirmed by such studies as those by Curry and Lillis (2004), Flowerdew (2007), and Belcher (2007).
8. Guillermo Paraje, Ritu Sadana, and Ghassan Karam, "Increasing International Gaps in Health-Related Publications," *Science*, May 13, 2005, 959–60; Phyllis Freeman and Anthony Robbins, "Editorial: The Publishing Gap between Rich and Poor; The Focus of AuthorAID," *Journal of Public Health Policy* 27 (2006): 196–203. Note this statement made by the latter authors: "A paucity of authors from developing countries in widely read and cited journals may help explain why global health policies tend to be determined with inadequate input from those with first hand experience and understanding" (197).
9. See for example Kerstin Stenius, Isidore Obot, Florence Kerr-Corea, Erickson F. Furtado, and Thomas F. Babor, "Reaching Parnassus: Advice on How to Get Published for Researchers from Developing or Non-English Speaking Countries," in *Publishing Addiction Science: A Guide for the Perplexed*, ed. Thomas F. Babor, Kerstin Stenius, and Susan Savva (London: International Society of Addiction Journal Editors, 2004), 33–44.
10. See for example Eren Zink, *Science in Vietnam: An Assessment of IFS Grants, Young*

Scientists, and the Research Environment, MESIA (Monitoring and Evaluation System for Impact Assessment) Impact Studies 9 (Stockholm: International Foundation for Science, 2009); http://www.ifs.se/Publications/Mesia/MESIA_9_Vietnam.pdf.

11. La Madeleine, "Lost in Translation," 454. The original specifies a "Japanese post-doc."
12. This is also mentioned as an important point by ibid., 455.
13. Thomson Reuters, "Science Citation Index Expanded," http://thomsonreuters.com/products_services/science/science_products/a-z/science_citation_index_expanded/ (accessed May 5, 2012).
14. See the extended discussion in Paul Wouters, *The Citation Culture* (Palo Alto, CA: Stanford University Press, 2003), especially chapter 1. Wouters points out (p. 4) that Garfield, a linguist by training, was inspired by earlier forms of citation analysis pursued by research librarians with limited budgets who were interested in finding out which journals were most important in different disciplines.
15. See for example a special 2010 issue of *Nature*, "Science Metrics," at http://www.nature.com/news/specials/metrics/index.html. Historically speaking, though reports such as the National Science Board's biannual Science and Engineering Indicators series had employed SCI data for decades, such measures only began to move out of the scholarly literature and into the policy arena from the late 1990s onward. An early paper that brought citation-based indicators to the attention of many in both the technical and the policy communities was Richard May's (humbly titled) "The Scientific Wealth of Nations," *Science* 275, no. 5301 (February 7, 1997): 793-96.
16. David A. King, "The Scientific Impact of Nations," *Nature* 430, no. 6997 (July 15, 2004): 311–16; quotations are from p. 311.
17. Shanghai Jiao Tong began creating its "academic ranking of world universities" in 2003. Four main criteria have been used: (1) "Quality of Education," measured by number of alumni who have won Nobel Prizes and Fields Medals (mathematics); (2) "Quality of Faculty," determined by existing faculty who have won these same prizes, plus highly cited researchers in twenty-one subject categories (predominantly scientific and engineering fields); (3) "Research Output," whose metric includes articles published in two journals, *Nature* and *Science*, as well as articles listed in SCI; and (4) "Size of Institution." Criteria 2 and 3 comprise 80% of the total weighting for rank determination.
18. Hannah Brown, "How Impact Factors Changed Medical Publishing—and Science," *British Medical Journal* 334 (March 17, 2007): 561–64.

19. Among the most prominent and influential global rankings are Academic Ranking of World Universities (formerly Shanghai Jiao Tong University ranking); Times Higher Education World University Rankings; QS (Quacquarelli Symonds) World University Rankings; Webometrics Ranking of World Universities; World Rankings by the Centre for Science and Technology Studies at Leiden University; and Newsweek International Rankings. In addition to these efforts, a number of ratings are focused specifically on scientific research, such as the Performance Ranking of Scientific Papers for World Universities, published since 2007 by the Higher Education Evaluation and Accreditation Council of Taiwan and the SCImago Institution Rankings produced and distributed since 2009 by the SCImago Research Group in Spain. Beyond these official ratings are dozens more that focus on research universities in specific regions (the EU, Middle East, Latin America, etc.) and individual countries. Nearly all these ratings employ citation data as well.
20. Eugene Garfield, "Citation Analysis as a Tool in Journal Evaluation," *Science*, November 7, 1972, 471–79. The author used the data reported to make a number of conclusions that upset many readers, causing more than a few to reject citation analysis and SCI data. An example of such a conclusion: "This analysis gives good reason or concern about any increase in the number of scientific and technical journals . . . so many journals now being published seem to play only a marginal role, if any, in the effective transfer of scientific information" (p. 474). Garfield, a tireless promoter of ISI (his own institute) and citation analysis generally (and well aware of its limitations, as the 1972 *Science* article shows), went so far as to lecture French scientists in 1976, in their own language, about their need to accept English as the international tongue of science. See Eugene Garfield, "La science française est-elle trop provincial?," *La Recherche* 7 (1976): 757–60.
21. Leo Egghe and Ronald Rousseau, *Introduction to Informetrics* (Amsterdam: Elsevier, 1990); Ulrich Ammon, *Ist Deutsche noch international Wissenschaftssprache? Englisch auch für die Lehre an den deutschsprachigen Hochschulen* (Berlin: de Gruyter, 1998); Thed N. van Leeuwen, Henk F. Moed, Robert J. W. Tijssen, Martijn S. Visser, and Anthony F. J. van Raan, "First Evidence of Serious Language-Bias in the Use of Citation Analysis for the Evaluation of National Science Systems," *Research Evaluation* 9 (August 2000): 155–56, and (same authors) "Language Biases in the Coverage of the *Science Citation Index* and Its Consequences for International Comparisons of National Research Performance," *Scientometrics* 51, no. 1 (2001): 335–46; Bo Sandelin and Nikias Sarafoglou, "Language and

Scientific Publication Statistics," *Language Problems and Language Planning* 28, no. 1 (2004): 1–10; and Hamel, "The Dominance of English in the International Scientific Periodical Literature."

22. Peder Olesen Larsen and Markus von Ins, "The Rate of Growth in Scientific Publication and the Decline in Coverage Provided by Science Citation Index," *Scientometrics* 84, no. 3 (September 2010): 575–603.
23. Sandelin and Sarafoglou, "Language and Scientific Publication Statistics," p. 4, table 1. The United Kingdom was twelfth and the United States was twenty-third.
24. Netherlands Observatory of Science and Technology, *Science and Technology Indicators 2010*, University of Leiden and Maastricht University (2011), summary in English; p. 28, table 4.3; http://nowt.merit.unu.edu/nieuwste_rapport.php (accessed July 2, 2012).
25. See for example Richard Smith, "Commentary: The Power of the Unrelenting Impact Factor—is It a Force for Good or Harm?," *International Journal of Epidemiology* 35, no. 5 (2006): 1129–30 and references therein, and Brian D. Cameron, "Trends in the Usage of ISI Bibliometric Data: Uses, Abuses, and Implications," *Portal: Libraries and the Academy* 5, no. 1 (January 2005): 105–25. For comparison, see also the article in favor of using bibliometric indicators to evaluate research quality: David Campbell (and 14 others), "Bibliometrics as a Performance Measurement Tool for Research Evaluation: The Case of Research Funded by the National Cancer Institute of Canada," *American Journal of Evaluation* 31, no. 1 (2010): 66–83.
26. Hartmut Haberland, "Domains and Domain Loss," in *The Consequences of Mobility: Linguistic and Sociocultural Contact Zones*, ed. Bent Preisler, Anne Fabricus, Hartmut Haberland, Sisanne Kjaerbeck, and Karen Risager (Roskilde, DN: Roskilde University, 2005), 227–37; http://magenta.ruc.dk/cuid/publikationer/publikationer/ mobility/mobility2/mobility_all/.
27. See for example Sirpa Leppänen and Tarja Nikula, "Diverse Uses of English in Finnish Society: Discourse-Pragmatic Insights into Media, Educational and Business Contexts," *Multilingua* 26, no. 4 (2007): 333–80.
28. NGU, Norges geologiske undersekellse; http://www.ngu.no/en-gb/ (accessed July 7, 2012).
29. Danish Agency for Science, Technology and Innovation, Ministry of Science, Innovation and Higher Education, Government of Denmark; http://en.fi.dk/ (accessed July 7, 2012).
30. David Bade, "Se vogliamo che tutto rimanga com'è, bisogna che tutto deva essere

scritto in inglese" (If We Want Things to Remain as They Are, Everything Must Be Written in English), *Biblioteche Oggi* 20, no. 4 (May 2012): 20–25.

31. Donald F. Thomson, "The Bindibu Expedition III," *Geographical Journal* 128, no. 3 (September 1962): 262–78; quotation is from p. 274.
32. Ibid., 274.

Chapter 5

1. Lynn Thorndike, *A History of Magic and Experimental Science* (New York: Columbia University Press, 1923), 2:20.
2. See for example Scott L. Montgomery, *Science in Translation: Movements of Knowledge through Cultures and Time* (Chicago: University of Chicago Press, 2002).
3. Nicholas Ostler, *Ad Infinitum: A Biography of Latin* (New York: Walker & Company, 2007). See also James Clackson and Geoffrey Horrocks, *The Blackwell History of the Latin Language* (Malden, MA: Blackwell, 2007).
4. The two standard references for this period are still those of G. E. R. Lloyd: *Early Greek Science: Thales to Aristotle* (New York: Norton, 1970) and *Greek Science after Aristotle* (New York: Norton, 1973). See also T. E. Rihll, *Greek Science* (Oxford: Oxford University Press, 2006); David C. Lindberg, *The Beginnings of Western Science: The European Scientific Tradition in Philosophical, Religious, and Institutional Context, Prehistory to A.D. 1450* (Chicago: University of Chicago Press, 2008); and James E. McClellan and Harold Dorn, *Science and Technology in World History* (Baltimore: Johns Hopkins University Press, 2006).
5. Lloyd, *Early Greek Science*; Otto Neugebauer, *The Exact Sciences in Antiquity* (New York: Dover, 1969).
6. Lloyd, *Greek Science after Aristotle*; McClellan and Dorn, *Science and Technology in World History*.
7. Good discussions of Strabo, Galen, and Ptolemy can be found in their respective entries in Charles Gillispie, ed., *Dictionary of Scientific Biography* (New York: Scribner's, 1970–80), and their updated version in Noretta Koertge, ed., *New Dictionary of Scientific Biography* (New York: Gale, 2007). On Ptolemy in particular, see for example Owen Gingrich, *The Eye of Heaven: Ptolemy, Copernicus, Kepler* (New York: Springer, 1997). For coverage of the Hellenistic age, see Anthony Kaldellis, *Hellenism in Byzantium: The Transformations of Greek Identity and the Reception of the Classical Tradition* (Cambridge: Cambridge University Press, 2008); Peter Green, *From Alexander to Actium: The Historical Evolution of the Hellenistic Age* (Berkeley:

University of California Press, 1993); and Geoffrey Horrocks, *Greek: A History of the Language and Its Speakers*, 2nd ed. (Oxford: Wiley-Blackwell, 2010).

8. Nicholas Ostler, *The Last Lingua Franca: English until the Return of Babel* (London: Walker & Company, 2010). See chapter 3, "The Pragmatism of Empire."
9. Gonzalo Rubio, "The Languages of the Ancient Near East," in *A Companion to the Ancient Near East*, ed. Daniel C. Snell (New York: Wiley-Blackwell, 2007); Horrocks, *Greek*.
10. Ostler, *Last Lingua Franca*, chapter 3.
11. For discussion of this crucial and often overlooked episode of transmission and preservation of Greek science, see De Lacy O'Leary, *How Greek Science Passed to the Arabs* (London: Routledge and Kegan Paul, 1949), and Montgomery, *Science in Translation*.
12. Ostler, *Ad Infinitum*. See also Clackson and Horrocks, *The Blackwell History of the Latin Language*.
13. On the handbook tradition in Hellenistic times, see W. H. Stahl, *Roman Science: Origins, Development, and Influence on the Later Middle Ages* (Madison: University of Wisconsin Press, 1962). On Aratus, see Emma Gee, *Ovid, Aratus and Augustus: Astronomy in Ovid's Fasti* (Cambridge: Cambridge University Press, 2009). Regarding Posidonius, see Ian G. Kidd and Ludwig Edelstein, eds., *Posidonius: The Fragments*, vol. 1 (Cambridge: Cambridge University Press, 1972).
14. For a discussion of Pliny's work, see John F. Healy, *Pliny the Elder on Science and Technology* (London: Oxford University Press, 1999). Online versions of the *Natural History* exist both in English, as part of the Perseus Project at Tufts University (http://www.perseus.tufts.edu/hopper/text?doc=Plin.+Nat.+toc [accessed July 7, 2012]; translation by John Bostock and H. T. Riley, 1855), and in Latin (http://penelope.uchicago.edu/Thayer/E/Roman/Texts/Pliny_the_Elder/home.html [accessed July 7, 2012]).
15. See John W. Humphrey, John P. Oleson, and Andrew N. Sherwood, *Greek and Roman Technology: A Sourcebook* (New York: Routledge, 1997), and J. G. Landels, *Engineering in the Ancient World*, rev. ed. (Berkeley: University of California Press, 2000).
16. See Ostler, *Ad Infinitum*, and Dag Norberg, *Manuel practique de latin medieval* (Paris: A. and J. Picard, 1968).
17. See Dmitri Gutas, *Greek Thought, Arabic Culture: The Graeco-Arabic Translation Movement in Baghdad and Early 'Abbasid Society* (London: Routledge, 1998), chapter 2.

18. See for example Hugh Kennedy, *When Baghdad Ruled the Muslim World* (Cambridge, MA: Da Capo Press, 2006), and, by the same author, *The Early Abbasid Caliphate* (London: Croom Helm, 1981), as well as I. M. Lapidus, *A History of Islamic Societies* (Cambridge: Cambridge University Press, 1988), p. 81, and, in general, Gaston Wiet, *Baghdad, Metropolis of the Abbasid Caliphate*, translated by Seymour Feiler (Norman: University of Oklahoma Press, 1979).
19. See Ostler, *Last Lingua Franca*, chapter 4, "When the Writ of Persian Ran."
20. Ibid. See also Montgomery, *Science in Translation*, chapter 3.
21. Gutas, *Greek Thought, Arabic Culture*, 5.
22. See Roshdi Rashed, ed., *Thabit ibn Qurra: Science and Philosophy in Ninth-Century Baghdad* (Amsterdam: De Gruyter, 2009).
23. See George Saliba, *Islamic Science and the Making of the European Renaissance* (Cambridge, MA: MIT Press, 2011), and A. I. Sabra, "The Appropriation and Subsequent Naturalization of Greek Sciences in Medieval Islam," *History of Science* 25, no. 60 (1987): 223–43.
24. Kees Versteegh, *The Arabic Language* (Edinburgh: Edinburgh University Press, 2001); see chapters 5 and 6.
25. See for example the classic discussion in Richard W. Southern, *The Making of the Middle Ages* (New Haven, CT: Yale University Press, 1961), 25–66, 74–110, and, more recently, Norman Cantor, *The Civilization of the Middle Ages* (New York: Harper, 1994).
26. Still very much worth reading on this topic is Charles Homer Haskins, *The Renaissance of the Twelfth Century* (Cambridge, MA: Harvard University Press, 1927). Competent surveys are offered by Lindberg, *Beginnings of Western Science*, and, more extensively, Montgomery, *Science in Translation*.
27. See for example Brian Stock, *The Implications of Literacy* (Princeton, NJ: Princeton University Press, 1987), and M. T. Clanchy, *From Memory to Written Record: England 1066–1307* (Oxford: Blackwell, 1993).
28. On these points, see Ostler, *Ad Infinitum*, 185–206.
29. For a detailed discussion, see Françoise Waquet, *Latin, or the Empire of a Sign: From the Sixteenth to the Twentieth Centuries*, translated by John Howe (London: Verso, 2003), chapters 3 and 4.
30. Ostler, *Ad Infinitum*, 293–94.
31. These and other data on Latin's fading glory are provided by Waquet, *Latin, or the Empire of a Sign*, 81–88, and Tim C. W. Blanning, *The Pursuit of Glory: Europe 1648–1815* (New York: Viking, 2007), 476.
32. See for example Daniel Kane, *The Chinese Language: Its History and Current Usage*

(North Clarendon, VT: Tuttle, 2006), and S. Robert Ramsay, *The Languages of China* (Princeton, NJ: Princeton University Press, 1989).

33. The most comprehensive source on Chinese science remains the series of volumes entitled *Science and Civilisation in China,* begun and mainly authored by Joseph Needham and published by Cambridge University Press. Recent volumes on specific topics in Chinese science have also been published by the Needham Research Institute and are listed on its website (http://www.nri.org.uk/otherworks.html). Essential information is also to be found in work by Nathan Sivin, including *Science in Ancient China: Researches and Reflections* (Aldershot, UK: Variorum, 1995); *Traditional Medicine in Contemporary China* (Ann Arbor: University of Michigan Center for Chinese Studies, 1987) (contains an extensive introduction to classical Chinese medicine); and writings and bibliography available on Sivin's website at http://ccat.sas.upenn.edu/~nsivin/index.html (accessed July 1, 2012). Also important are works by Shigeru Nakayama, such as *A History of Japanese Astronomy: Chinese Background and Western Impact* (Cambridge, MA: Harvard-Yenching Institute, 1969) and *Academic and Scientific Traditions in China, Japan, and the West* (Tokyo: University of Tokyo Press, 1984), as well as relevant discussion in Toby E. Huff, *The Rise of Early Modern Science: Islam, China, and the West* (Cambridge: Cambridge University Press, 2003), 240–324.
34. Nakayama, *History of Japanese Astronomy*, 9–10.
35. John K. Fairbank and Merle Goldman, *China: A New History*, 2nd ed. (Cambridge, MA: Belknap Press of Harvard University Press, 2006), 72–83.
36. Masayoshi Sugimoto and David L. Swain, *Science and Culture in Traditional Japan* (North Clarendon, VT: Tuttle, 1989), 16–17.
37. Nakayama, *History of Japanese Astronomy*, 11.
38. Dieter Kuhn, *The Age of Confucian Rule: The Song Transformation of China* (Cambridge, MA: Belknap Press of Harvard University Press, 2009). See also Asaf Goldschmidt, *The Evolution of Chinese Medicine: Song Dynasty, 960–1200* (London: Routledge, 2008).
39. Goldschmidt, *Evolution of Chinese Medicine*; see especially chapter 4.
40. Sivin, *Science in Ancient China.*
41. Mark Elvin, *The Pattern of the Chinese Past* (Stanford, CA: Stanford University Press, 1973), 114–94.
42. Heyem Tongming, "Rice Cultures South of Yangtze River and Japan," *Agricultural Archaeology* 1 (1998): 335–43.
43. See Kuhn, *Age of Confucian Rule*, 160–86.
44. See the summary discussion by Huff, *Rise of Early Modern Science*, 242ff.

45. See Nakayama, *History of Japanese Astronomy*, chapter 9.
46. Ibid.
47. Ostler, *Last Lingua Franca*. See chapter 12, "Under an English Sun, the Shadows Lengthen."
48. See for example Yorick Wilks, *Machine Translation: Its Scope and Limits* (New York: Springer, 2008), especially chapter 2.

Chapter 6

1. "Revised Rules for Botanical Taxonomy Take Effect," newsblog Nature.com, January 9, 2012; http://blogs.nature.com/news/2012/01/revised-rules-for-botanical-taxonomy-take-effect.html (accessed June 19, 2012).
2. Christian Mair, *Twentieth-Century English: History, Variation, and Standardization* (Cambridge: Cambridge University Press, 2006), 193.
3. Ibid.
4. The following points are based on conversations and interviews with teachers, students, and adults from each of the noted countries (the Netherlands, Sweden, Denmark, Norway, Finland, and Iceland), as well as various readings that include the following: Pasi Sahlberg, *Finnish Lessons: What Can the World Learn from Educational Change in Finland?* (New York: Teachers College Press, 2011); Jim Allen, Yuki Ineaga, Rolf van den Velden, and Keiichi Yoshimoto, eds., *Competencies, Higher Education and Career in Japan and the Netherlands* (New York: Springer, 2010); Johanna Einarsdottir, ed., *Nordic Childhoods and Early Education: Philosophy, Research, Policy and Practice in Denmark, Finland, Iceland, Norway, and Sweden* (Charlotte, NC: Information Age Publishing, 2006); Hans Peter Jensen and Henrik Johannesson, "Engineering Courses Taught in English: An Example from Denmark," *European Journal of Engineering Education* 20, no. 1 (1995): 19–23; Sirkku Latomaa and Pirkko Nuolijarvi, "The Language Situation in Finland," in *Language Planning and Policy in Europe*, vol. 1: *Hungary, Finland and Sweden*, ed. Robert B. Kaplan and Richard B. Baldauf Jr. (Bristol: Multilingual Matters, 2005), 125–232; Evanthia K. Schmidt, "Higher Education in Scandinavia," in *International Handbook of Higher Education*, ed. James Forest and Philip Altbach (Berlin: Springer, 2011), 18:977–96; Karen Sundin, "English as a First Foreign Language for Young Learners: Sweden," In *An Early Start: Young Learners and Modern Languages in Europe and Beyond*, ed. M. Nikolov and H. Curtain (Strasbourg: Council of Europe, 2000), 151–58; L. K. Sylvén, "How Is Extramural Exposure to English among Swedish School Students Used in

the CLIL Classroom?," *Vienna English Working Papers* 15, no. 3 (2006): 47–54; http://ec.europa.eu/education/languages/eu-language-policy/index_en.htm; Susan Wilborg, *Education and Social Integration: Comprehensive Schooling in Europe* (New York: Palgrave Macmillan, 2009); and Birger Winsa, "Language Planning in Sweden," in *Language Planning and Policy: Europe; vol. 1, Hungary, Finland, and Sweden*, ed. R. B. Kaplan and R. B. Baldauf Jr. (Bristol: Multilingual Matters, 2005), 233–330.

5. Dutch Eurydice Unit, *The Education System in the Netherlands*, Ministry of Education, Culture and Science, November 2007; online through the ministry website at http://english.minocw.nl/documenten/en_2006_2007.pdf.
6. *From Linguistic Diversity to Plurilingual Education: Guide for the Development of Language Education Policies in Europe*, Language Policy Division, Council of Europe, 2007, 36; http://www.coe.int/t/dg4/linguistic/Publications_EN.asp#P48_1276.

INDEX

‘Abbasid dynasty, 142
Adelard of Bath, 132–33, 134, 146, 147, 157, 159
Africa, 4, 13, 15, 24, 27, 38, 41, 42, 44, 76, 92, 129, 175; Arabic in, 44, 51, 145; East Africa, 1, 14, 43; English, varieties of in, 14, 15, 58–59; French, decline of in, 46; North Africa, 11, 12, 29, 36, 44, 86, 133, 135; science, future of in, 162–63; scientific publication in English and, 86; sub-Sahara, 77; West Africa, 14, 15, 16, 33, 94
Afrikaans, 55
Agency for Science, Technology and Innovation (Denmark), 125–26
Akasaka, Kiyo, 48
Alfonso VI of Castile, 145
Algeria, 12
Al-Minya University, 167
Amharic, 24
Andalusia, 144
Antarctica, 28
Arabic language, xi, 13, 22, 24, 35, 97; as international language, 11, 26, 36, 44, 45 table 2.1, 61, 123, 147–48, 157, 171; impact of, on other tongues, 51, 143; Internet usage of, 47; as past lingua franca of science, 8, 104, 128, 133, 134, 135, 141–46, 150, 154, 158, 159, 163, 178; scientific publication in, 4, 126
Arabic numerals, 144
Aramaic, 137, 138, 142, 143
Argentina, 13, 77, 128
Aristotle, 136, 138, 139, 140, 143, 144
Arnold, Mathew, 150
Assiut University, 166
Association of Southeast Asian Nations (ASEAN), 40
Atlas of the World’s Languages in Danger, 50

Australia, xii, 33, 52, 81, 102, 167; endangered languages in, 50, 51; English and, 58; international students in, 31; scientific publication in, 107
Austria, 94
Averroes, 144

Bacon, Francis, 148–49, 154, 169
Baghdad, xi–xii, 10, 142, 144, 145
Bangladesh, 14, 28, 46
Bayt al-Hikma, 10
Belarus, 78
Belgium, 30, 42, 43, 121
Bengali, 45 table 2.1, 46, 53, 56
Berber language, 143
Bernoulli, Daniel, 71
Berthollet, Claude Louis, 71
bibliometrics, 83, 119
Bible, 21, 49fn
big science, 11, 87, 112
Bindibu people, 130–31
Black, Joseph, 71
Botswana, 78
Boyle, Robert, 149
brain drain, 70, 78–81
Brazil, 2, 3, 15–16, 31, 35, 50; economy in, 16; educational laws in, 17; endangered languages in, 50; English-language programs in, 28; English use in, 9, 35, 36, 127; R&D spending by, 76; researchers increasing in, 77; science in, 16–17, 79, 80, 112, 176; scientific publication in, 83, 86, 99, 170
Bretton Woods Agreement, 34
BRIC (Brazil, Russia, India, China), 16–17
Buffon, Georges-Louis Leclerc, 71
Bulgaria, 94, 98
Bush, Vannevar, 74, 100, 176; "Science—The Endless Frontier," 100

Cairo University, 167
Cambodia, 115
Cameroon, 43
Canada, 14, 33, 43, 79; English use in, 58, 73; international students in, 30, 31; language endangerment in, 50, 51; scientific publication in, 107, 121
Cantonese language, 6, 46
Cassini, Jean-Dominique, 71
Catalan language, 147
Catherine the Great, 68
CERN (Conseil Européen pour la Recherche Nucléaire), 4
Cervantes, Miguel de, 149
Chad, 38
Chartres Cathedral, 146
ChemWeb, 4
Chernobyl, 69
China, 78, 86; economic development and science in, 16–17, 39, 53, 66; English learning in, 13, 18; Internet in, 46; languages in, 46; Project 985, 17; R&D capability of, 17, 79, 112, 170, 176; researchers increasing in, 77; science and technology in, history, 125, 142, 150–54, 211n33; scientific advancement in, 17, 66, 84; scientific collaboration and, 80; and students studying abroad, 5–6, 31, 66, 79; university training in science and engineering and, 16–17
Chinese Academy of Sciences, 121

Chinese language (Mandarin), 15, 18, 44, 46; as competitor to English, 5–7, 22, 48, 84, 161; importance of, for non-Chinese scientists, 173, 185–86; as international language, 104, 110, 113, 123, 171; Internet usage of, 47; and journals switching to English, 6, 84, 100; as past lingua franca of science, 134, 150–54, 157; scientific publication in, 4, 20, 100, 108, 109, 121, 189n4; speakers of, 5, 45 table 2.1, 46
Chinese Ministry of Education, 32
Chinese people: number of, studying English, 6–7; and scientific papers published in English, 5, 84, 94, 98; and STEM researchers, 84
Cicero (Marcus Tullio Cicero), 148
circles of English. *See* English, circles of
citation analysis, 205n14, 206n20
coal, 16
Cold War, 16, 31, 34, 79
collaboration, 2, 77, 79, 92, 111, 129, 164; anglophone countries and, 126, 129; growth of, in science, 80, 86–89, 124, 173, 175, 185; as primary dimension to scientific work, 87–89; quantified levels of, 94, 95 fig. 3.6, 118
colonialism, 19, 29, 44, 186; and Chinese, in Korea, 151; and classical Greek, 135–36; and English, 55, 56–57, 61, 169; impact of, on languages, 44, 49, 51, 55, 64, 113
Content and Language Integrated Learning (CLIL), 182–83, 184, 185
Copernicus, Nicolas, 148
Coptic language, 143
Cordoba, 8
Council of Europe, 181, 186
Council of Science Editors, 60
creole (language category), 26
Crusades, 145
Crystal, David, iv, vii, x, xiii, 25, 26, 27, 29, 45 table 2.1, 190n11, 196n37
Cyrillic, 49
Czechoslovakia (former), 78
Czech Republic, 30, 88, 94

d'Alembert, Jean le Rond, 71
Darwin, Charles, 15, 19, 72, 134, 166
Denmark, 43, 77; English-language programs in, 181–82; English in school curriculum of, 181–82, 212n4; English use in, 30, 36, 123, 125; impact factor scores in, 121; language policy in, 124; scientific publication in, 94, 121
Descartes, René, 21, 148, 149, 154
domain loss, 123–25, 184

East African Rift System, 43
East Asia, 11, 34, 48, 63, 73, 94, 117, 121, 173; Chinese as lingua franca in, 150–54; and cultural aspects of expression in science, 186; English learning in, 19, 28; and plagiarism in published science, 108; R&D spending by, 76
l'École Polytechnique, 10
Egypt, 28, 97, 98, 125, 166, 178; ancient, 136–37, 140, 172
Ehrlich, Paul, 75

Einstein, Albert, 72, 101, 134, 187
England, 25, 28, 33, 57, 65, 73, 132–33, 147, 155, 171; and boycott against German scientists, 75; colonial language policy of, 33; and Latin books published in 1700, 149; researchers increasing in, 77; scientific development in, 70, 72; scientific publication in, 88, 92
English: as basic skill, 26, 28, 110; circles of (Kachru), 61–63, 64, 65, 98, 99, 163, 171, 183, 197n47; as colonial language, 22, 29, 33–34; as dominant foreign language in schools, 29–30; as foreign language (EFL), 7, 61, 63, 107–8, 110, 111, 179, 180, 183; geopolitical aspects of, 12, 16, 19, 37–41, 66, 94, 169; and global diplomacy, 40; as global language, ix, 3, 8, 12, 24–68, 168, 174; impact of, on indigenous languages, 50–51; Internet usage of, 46–49; as "killer" language (of other tongues), 9, 52; as the language of science, ix, xii, 3–4, 7, 8, 11, 68–102, 105, 163, 166–87; as the language of university instruction, 6, 32–33; as native language (ENL), 61, 63; nonstandard (in science), 129, 163, 178; number of speakers of, 3, 45 table 2.1; as official language, 28, 41; rise of, factors leading to, 12–13, 37–41; in Rwanda, 41–44; as second language (ESL), 7, 14, 61, 63; Standard (written) (SE, SWE), 14, 57–60, 98; teaching and learning of, 19, 65, 174, 179–85
English as a Global Language (Crystal), ix, 25, 45 table 2.1, 190n11, 192n2, 193n7
Esperanto, 21, 162
Estonia, 113
ethanol, 15, 16
Ethiopia, 24–25, 43, 64
Ethnologue, 27, 45 table 2.1, 49, 50, 192n4
Euclid, 133, 136, 153
Euler, Leonhard, 71
European Science Foundation, 4
European Union (EU), 37, 41, 58, 84, 121, 123, 183, 206n19; language policy of, 36

al-Farabi, 144
Finland, 6; English in school curriculum of, 182, 212n4; English-language programs in, 181–82; English use in, 30, 117, 124; scientific publication in, 121
Fischer, Emil, 126
Fong, Shih Choon, 10
France, 10, 13, 33, 36, 74, 75, 133, 159, 180; English learning in, 117; English use in, 30; as European power, 70; language policy in, 51, 148; and Latin books published in 1700, 149; and Rwanda genocide, 42; scientific publication in, 72, 88, 94, 120
Franklin, Benjamin, 71
French language, 12, 15, 22, 46, 133, 147, 187; as colonial language, 41,

42, 113; as international language, 26, 29, 44, 46, 113, 126, 181; as language of science, 4, 11, 70, 72, 73, 89, 92, 93 fig. 3.5, 108, 115, 134, 149, 173, 185; as past lingua franca in Europe, 18–19, 72, 132, 149; as vernacular, 147, 154
The Future of English (Graddol), 26–7, 192n3

Galen (Claudius Galenus), 136, 144, 208n7
Galileo Galilei, 148, 149, 154, 187
Gambia, 14
Gandhi, Mohandas, 57
Garfield, Eugene, 106–7, 118–19, 205n14, 206n20
Geike, Archibold, 155
Geim, Andre, 68–70, 75, 77–78, 80, 81, 88, 89–90, 101, 180, 197n1
Geological Survey of Norway, 124
Geologisches Rundschau. See *International Journal of Earth Sciences*
geopolitics, xii, 12, 16, 19
GeoRef, 4
Gerard of Cremona, 146
German language, 12, 45 table 2.1, 71, 72, 147, 154, 187; boycott against scientific papers and presentations in, 73, 75, 94; as international language, 22, 29, 44, 46, 73, 126, 161, 181; as past lingua franca of science, 11, 65fn1, 73, 89, 92, 93 fig. 3.5, 115, 134, 155, 173
Germany, 10, 37, 69, 148, 161; ; English use in, 30, 65, 116, 117; as scientific power in modern Europe, 73, 74, 156, 176, 187; trends in scientific publishing in English and, 88, 92, 120, 155
Ghana, 14
Glasgow University, 149
global language, xii, 3, 8, 101; advantages and disadvantages of, 102–31; geopolitical aspects to, xii, 34; impacts of, in research, 34, 63, 170, 171, 174, 176. *See also* English, as a global language
globalization, 3, 16, 41; criticisms of, 52, 171; English as medium of, 9, 22, 52, 107; international students and, 31–32; of science, 23, 77, 86, 88, 101, 104, 111, 125; and spread of English, 13, 22, 66; US role in, 12, 22, 34
Google Scholar, 82, 83
Graddol, David, 25, 26, 27, 45 table 2.1, 110
graphene, 68, 69, 78, 101
Great Sandy Desert, 130
Greek (ancient), 61, 135, 142, 144, 145, 146, 159, 160, 178; as past lingua franca of science, 104, 134, 135–38, 139, 140, 143, 147, 154, 171
Guam, 38

Haeckel, Ernst, 75
Haiti, 38
Hakluyt, Richard, 149
Halley, Edmund, 149
Han dynasty, 151
Hanzi (Chinese characters), 150

Harran (city), 142, 144, 158
Helwan University, 167
Hezbollah, 38
Hindi, 45 table 2.1, 46, 55, 56
Hitler, Adolf, 73, 187
HIV/AIDS, 39
Hong Kong, 31, 55; variety of English in, 14, 56
Hooke, Robert, 149
Horace (Quintus Horatius Flaccus), 148
Hugh of Santalla, 146
Hunayn Ibn Ishaq, 144, 146
Hutcheson, Francis, 149
Huxley, Thomas, 155
Huygens, Christian, 71, 149

Ibn al-Haytham, 144
Ibn Sina (Avicenna), 144
Iceland, 30, 181, 182, 212n4; scientific publication in, 121
impact factor (IF), 83, 118, 119, 120, 122, 163
India, 2, 10, 21, 36, 40, 61, 78, 86, 180; conflict with Pakistan, 38; economic development and science in, 16–18, 39, 41, 175; English use in, 56–57, 63, 99; Internet in, 46; language endangerment in, 52; languages in, 27, 46, 55; R&D capability of, 76, 79, 112, 170, 176; researchers increasing in, 77; science and technology in, history, 125, 142, 145; scientific collaboration and, 80, 88; and students studying abroad, 5, 31; variety of English in, 14, 58–59
Indonesia, 13, 28, 39, 94
Industrial Revolution, 22, 33, 37, 72, 73, 149
Institute of International Education, 30
Institute for Scientific Information (ISI), 82, 106, 119, 120
International Atomic Energy Agency, 41
International Criminal Court, 41
International Energy Agency, 40
International Journal of Earth Sciences (Geologisches Rundschau), 91–96, 155
International Monetary Fund, 13, 42
international students, xi, 5, 10, 17, 30–32, 79
Internet, 4; English as dominant language of, 46–49, 120; globally restricted access to, 47–48; and growth of English, 13, 46–49, 161; impact on modern science, xii, 135; languages on, 46–48, 171; use of, in classroom, 181
Iran, xi, 28, 38, 78, 86, 137
Iraq, xi, 38, 40, 138
Ireland, 36, 58, 121
Islam, 11, 22, 171; "golden age" of, 10; and Mongols, xii, 152; and science, 10, 125, 139, 141–45, 146, 147, 150, 153, 158
Israel, 38, 70, 94

Jamaica, variety of English in, 14, 56
Japan, 2, 3, 12, 31, 36, 37, 48, 74, 81, 100, 180; Chinese as past lingua franca in, 123, 128, 150, 151, 153–54, 157; and citation data use, 119, 120; earthquake of 2011, 39; English and, 18, 28, 30, 32, 40, 48, 52, 75,

116, 117; English teaching/learning in, 18, 28, 52, 127; nuclear politics and, 38; R&D capability of, 17, 39, 100, 112; science in, 39, 73, 75, 76, 150–51, 152, 154, 157, 175, 176; scientific publication in, 86, 89, 115fn1, 120, 173
Japanese language, 18, 65, 107fn1, 185; Chinese influence on, 150, 151, 157; as colonial language, 54, 113; science in, 4, 108, 154, 173; scientific publication in English and, 89; speakers of, 45 table 2.1, 46
Jiang Zemin, 17
John of Seville, 146
Jordan, 78
Journal des sçavans, 70, 71

Kachru, Braj, 61, 197n47; and circles of English, 61, 62 fig. 2.1, 63, 191n17
Kagame, Paul, 42
Kazakhstan, 78
Kenya, 14, 42, 54
al-Khwarizmi, 133, 144
al-Kindi, 144
King Abdullah University of Science and Technology (KAUST), 9–11
Kinyarwanda language, 41, 42, 44
Korea, 28, 31, 35, 38, 39, 54, 74, 81, 151, 152; Chinese as past lingua franca in, 123, 150, 151, 153–54; and citation data use, 119; English teaching/learning in, 18, 28, 127; English use in, 28, 30, 40, 116; R&D capability of, 17, 79, 112; scientific publication in English and, 75, 86, 98
Korean language, 4; influence of Chinese on, 150
Kriol, 103
Kublai Khan, 152

Lake Kivu (Rwanda), 41, 42, 43
Lake Nyos, 43
language: and children, 34, 42, 49–50, 52, 53, 67, 148, 180; and (linguistic) diversity, 9, 26, 54, 122, 125; and (linguistic) genocide, 52; global, 34, 63, 101, 102–32, 174; learning of, 63–64, 127–28, 157, 161; universal, 3, 21, 26, 36, 82, 154, 156
language death, 52, 123, 134, 171
language endangerment (threat), 49–53
language/linguistic injustice (inequality), 107–11, 203n2
language policy, 33, 52, 56, 124, 202n2
language preservation, 50–51
language/linguistic rights, 53, 124
Laplace, Pierre-Simon, 71
Large Hadron Collider, xii, 87, 173
The Last Lingua Franca (Ostler), 209n8
Latin, 8, 61, 70, 72, 81, 133, 210n31; Galileo and, 187; as past lingua franca of science, 8, 74, 104, 133–34, 138–41, 145–50, 154, 156, 157, 158, 169
Latin America, 16, 19, 76, 206n19; English use in, 13, 20, 29, 117, 129; and language, 5, 36, 51; and modern science, 162; and scientific journals, 83; researchers increasing in, 77; scientific publication in, 86, 94, 120
Lavoisier, Antoine, 71
Lebanon, 38

Leibnitz, Gottfried, 71
Liberia, 14, 55
Lichtenstein, 46
lingua franca, viii, x, 22, 103, 104, 177; ancient Greek as, 135–38; Arabic as, 11, 142–45, 158; Chinese as, 150–54; English as, 25, 28, 40, 61, 75, 116, 160, 164; French as, 46, 70; German as, 72, 73, 75; impacts of, on other languages, 156–57, 159; Kriol as (Australia), 103; Latin as, 8, 74, 138–41, 145–50, 169, 187; relationship of, to native-speaking country; 156–57, 159, 170; Russian as, 40; in science, x, 22, 95, 98, 104, 110, 125, 128, 132–66, 177, 179
linguicide, 9
linguistic imperialism, 52, 190n9
Linnaeus (Carl Nilsson Linneaus), 71
Li Shizen, 153
Lucretius (Titus Lucretius Carus), 139
Luxembourg, 46

Macaulay, Lord Thomas Babington, 56, 64
Macaulay Minute, 56, 197n42
Madrid, 32, 38
Makerere University, 58
Malaysia, 31, 39, 78; language policy of, 128; scientific publication in English and, 86; variety of English in, 56
al-Mamun, 142
Manifesto of the Ninety-Three, 75
al-Mansur, 142
Martianus Capella, 140, 141
Maupertuis, Pierre, 71
Max Planck Institute, 4
mercantilism, 78fn
Mercator, Gerardus, 71
Mersenne, Marin, 21
Mesopotamia, 144, 172
Mexico, 113, 127, 173, 175
Ming dynasty, 153
MIT (Massachusetts Institute of Technology), 10
Moldova, 78
Mongols, xii, 152–53
Montaigne, Michel de, 149
Moscow Engineering Physics Institute, 69
Moscow Institute of Physics and Technology, 69
Mozambique, 81
Muhammad (Prophet), 11
multilingualism, 26, 34, 54, 93, 157, 161, 162, 183
Mumbai, 38
Myanmar, 38

Namibia, 14, 78
Napoleon (Napoleon Bonaparte), 19
National Education Guidelines and Framework Law (Brazil), 17
nationalism, 4, 57, 79, 134, 150
National Science Board, 76, 205n15
National University of Singapore, 10
Nature, 96, 119, 205n15, 205n17; first issue of, 155
Nazi, 73, 161
Nebrija Universidad, 32
Netherlands, 69, 74; English in school curriculum of, 65, 110, 182, 212n4; English use in, 30, 116, 123, 125,

127; English-language programs in, 117, 180, 181–82; impact factor scores for, 121; language policy in, 124; scientific publication in English and, 88, 94, 121
new Englishes, x, 13–14. *See also* World Englishes
Newton, Isaac, 72, 149, 154
New York City, 53, 59
New Zealand, 30, 33, 58, 77
Nigeria, 28, 34; variety of English in, 14, 55
Nobel Prize, 68, 101, 205n17
North Korea, 37
Norway, 2, 123; English in school curriculum of, 182, 212n4; English use in, 30, 117, 124, 125, 127; English-language programs in, 181–82; impact factor scores for, 121; scientific publication in, 121
Nottingham University, 69
Nye, Joseph, 29

OBIS (Ocean Biogeographic Information System), 4
Oestler, Nicholas, 161
Office of Scientific Research and Development (USA), 100
Olympics, 36
Organization for Economic Cooperation and Development (OECD), 30, 39, 76
Ørsted Institute, 69

Pahlavi language, 137, 142, 143, 145
Pakistan, 11, 14, 28, 38, 40, 137
Panama, 64
Papua New Guinea, 50
Paris, 8, 19, 40
Persia, 10, 123, 128, 137, 138, 142, 143, 144, 153
Persian language, 61, 134, 142, 143, 152; as regional lingua franca, 137, 171
Petrarch (Francesco Petrarca), 148
Phags-pa script, 152
Philosophical Transactions of the Royal Society of London, 70, 71
pidgin, xi, 26, 103
plagiarism, 100, 108, 179
Planck, Max, 75
Plato of Tivoli, 146
Pliny the Elder, 139, 141, 209n14
Poland, 30, 94, 113, 128
Portuguese, 4, 44, 45 table 2.1, 47, 53, 83, 126, 185
Posidonius (of Apamea), 136, 139, 159, 209n13
Priestley, Joseph, 71
Ptolemy (Claudius Ptolemy), 136, 137, 138, 140, 143, 144, 208n7
PubMed, 4, 201n34

Qur'an, 142, 147

Rabelais, Françoise, 149
Radboud University, 69
al-Razi, 144
Red Sea, 9, 10, 97, 166, 168
research and development (R&D), 76, 88, 175; in BRIC nations, 17, 176; in China, 17, 79, 84, 170; in private sector, 4, 74, 77, 115, 134, 135, 170, 173; as scientific capacity building, 79, 84, 112, 170, 173, 176; as source

research and development (R&D) (*continued*)
of economic development, 76, 116, 176; spending on, 76–77, 79, 83, 84; and spread of English, 4, 115, 170, 173; in United States, 74; using publication output to evaluate, 118–19
Ricci, Matteo, 153, 154
Roentgen, Wilhelm, 75
Romania, 78
Royal Society of London, 70, 71, 79, 85 fig. 3.1, 87 fig. 3.2, 149
Rozier, François, 71–72, 75, 104, 154
Russia, 10, 13, 16, 69, 81, 180; and brain drain, 78; ; English learning in, 28, 127; international students from, 31; R&D capability of, 112; scientific collaboration and, 94: and spread of English, 38, 39–40
Russian language, 26, 29, 40, 45 table 2.1, 46, 68, 89, 113; as international language, 22, 45 table 2.1, 113, 126, 127, 173, 185; Internet usage of, 47; and language endangerment, 51, 53; scientific publication in, 4, 11, 89, 92, 93, 107fn1, 173
Rwanda, 38, 41–44, 61, 63; English use in, 43–44; genocide in, 41–42; language policy of, 41
Rwandan Patriotic Front (RPF), 41, 42

Saddam Hussein, 38
Samarkand, 142
Sanskrit, 134, 142
Sapir-Whorf hypothesis, 125
Saudi Arabia, 9, 10, 99
Scheele, Carl Wilhelm, 71
Science, 96, 205n17
Science Advisory Council (India), 17
Science Citation Index (SCI), 82–83, 89, 106, 118–20, 122
"Science—The Endless Frontier" (Bush), 100
Scientific Revolution, 21, 33, 37, 72, 145, 149
Scopus, 82, 83, 87, 106, 119, 122
Selassie, Haile, 24
Seung-Hwan Ham, 28, 29, 90
Shanghai, 20
Shanghai Jiao Tong University, 119, 205n17, 206n19
Shen Kua, 152
Siberia, 50, 51, 52
Sicily, 133, 135, 145, 146
Singapore, 10, 18, 63; higher education in, 30, 31; scientific achievement in, 10, 86; variety of English in, 14, 34, 56
Six Party Talks, 38
Smith, Adam, 24, 78, 149
Sochi, 68
soft power, 160
Sogdian, 142, 143
Somalia, 38
Song dynasty, 151–52, 153, 159
Song Yingxing, 153
South Africa, 14, 30, 37, 42, 43, 78; linguistic policy of, 54–55; R&D capability of, 112–13
Soviet Union, 34, 37, 46, 69, 73, 94, 120, 173
Spenser, Edmund, 149
Sri Lanka, 14, 63
Stalin, Joseph, 68